FLYING ROAST DUCKS

RECOLLECTIONS OF SIR HERMANN BONDI
1983-2005

PAULA HALSON

"Man who waits for roast duck to fly into mouth must wait very, very long time"

(Chinese proverb)

'Flying Roast Ducks' was one of the five suggestions put forward by Hermann Bondi as a possible title for his autobiography, eventually entitled *Science, Churchill & Me*

Published by Churchill College

Published by Churchill College, Cambridge © 2012

ISBN: 978-0-9563917-3-5

Not to be reproduced or transmitted in any form without permission from the publisher

Editor: Tim Cribb

Written by: Paula Halson

Cover photography: Gavin Bateman

Design: cantellday – www.cantellday.co.uk

Printer: Norwich Colour Print

Photographs: James Adamcheski-Halson (p. 95), American Institute of Physics [copyright uncertain], (p. 11), Gavin Bateman (front cover, pp. 6, 88), Christine Bondi (pp. 8, 15, 31-32, 34-35, 38-39, 41-42, 46, 48, 52, 55, 57, 59, 68, 84, 98, 102, 105), Sid Brown (p. 50), Cambridge Evening News (p. 26), Tim Cooper (pp. 24, 63-64), Paula Halson (pp. 4, 18, 21, 36, 40, 60, 70, 90, 96), Joanna Hawthorne-Amick (p. 44), Mike Laycock (p. 51), John Edward Leigh (p. 16), June Mendoza (p. 87), Dame Kathleen Ollerenshaw (p. 79), Barry Phipps (back cover). Images at pp. 16, 23, 26-27 and 76 are held by the Churchill Archives Centre.

Typeset in Gill Sans

For further information on the Bondi Papers, please contact:

Churchill Archives Centre

Churchill College

Cambridge

CB3 0DS

Tel: 01223 336087

www.chu.cam.ac.uk/archives

Front cover: Name detail on the door of Room 50A. Back cover: The Master's Lodge (2011)

PREFACE

I was Sir Hermann Bondi's secretary from 1988-90 and was appointed Registrar in 1991, a post I still hold. Following Sir Hermann's death in 2005 I was asked by the Director of the Archives Centre, Allen Packwood, to organise the files that remained in Sir Hermann's College room in preparation for their transferral to the Archives Centre. This memoir is the unexpected product of that request. It is a brief biography based on personal knowledge and reminiscences.

I did not, for one minute, set out to write a memoir of Sir Hermann Bondi, or in fact to write anything at all. I am an administrator. I write minutes and policy documents. That is where my expertise lies. Yet as I sorted the papers in Sir Hermann's College room, I could see that they contained snippets of information and stories of life away from the public image that could paint a more colourful picture of Sir Hermann's life. However, the reader should know that I make no attempt to examine Sir Hermann's scientific achievements or debate his views on education or religion, not least because I am patently unqualified to do so. In any case, Sir Hermann's many great scientific and administrative achievements are already known from the literature, such as the biography written by a Churchill graduate, Professor Ian Roxburgh,[1] or the expanded version of the talk given by Mark Goldie, Fellow and former Vice-Master, at Sir Hermann's Memorial Concert on 26 November 2005.[2]

Although I will be turning to some of the more prosaic events such as problems with the Master's Lodge heating system, parking on the forecourt, or even the difficulties of managing the Master's diary, I will also be looking at the more interesting, including an unusual view of Magic Squares and an intriguing exchange between two highly enthusiastic protagonists. To this, I have added my own recollections, those of the Bondi family, members of the College and other acquaintances. In this way, I also hope to offer a glimpse into the life of a former Master of the College and, in so doing, to provide an opportunity to view Churchill College at an operational level. A list of contributors and an index of names included in this memoir can be found in the appendices.

Sir Hermann married Christine Stockman on 1 November 1947 and they were to form a very special partnership that was to endure nearly sixty years until his death on 10 September 2005. The importance of the role played by Christine is ever present. She was to provide the bedrock for every project or activity Sir Hermann undertook. Notwithstanding the demands of bringing up five children, they co-wrote papers and articles together and he encouraged her in her academic and other pursuits. The very special relationship that the College continues to enjoy with Sir Winston's daughter, The Lady Soames, DBE, is due in no small part to Sir Hermann and Lady Bondi's friendship and warmth and their deep commitment to the College.

Sir Hermann's autobiography, *Science, Churchill & Me*, was published on 10 May 1990 to celebrate the 50th anniversary of Sir Winston Churchill's appointment as Prime Minister. There is reference in this memoir to a school's marking of the 50th anniversary of the declaration of the Second World War, while I comment on a dinner party held on 20 September 1987 to celebrate the 50th anniversary of Sir Hermann's arrival in the country. I am pleased that this memoir of Sir Hermann seems to be part of this recurring theme, having been written in the year that Churchill College celebrates its 50th anniversary.

I would like to make one final point. With the exception of the opening biographical chapter, throughout I repeatedly refer to Sir Hermann as 'Sir Hermann', rather than 'Hermann'. This is a reflection of my professional training and I would have found it difficult to have addressed him otherwise although I know that he would have been more than happy for me to have addressed him as 'Hermann'. The fact that he did not mind what he was called is to me a measure of how approachable he was. If he was alive today I am sure he would be chuckling about it now.

PAULA HALSON

Sir Hermann and Paula Halson at a staff party (2001)

FOREWORD

What does the Master actually do? and Are you enjoying it? are probably the two questions most frequently asked of a Head of College in Oxford or Cambridge. The public answer to the latter is invariably Yes; the longer responses to both are a good deal more subtle and informative.

Sir Hermann Bondi was the third Master of Churchill College, Cambridge, following a most distinguished career in science and scientific leadership (I prefer to use that word rather then administration!). Paula Halson was his Personal Assistant during his time as Master and was also responsible for the organisation of his papers before they were deposited in the Churchill Archives Centre.

Paula therefore saw the man and the job in a way that very few could and her memoir gives revealing and delightful insights into the ways of Churchill College and into Hermann himself. Many personal qualities and characteristics of him shine through, above all his pleasure (almost all the time…) in engaging with people.

And that would be my short answer to both questions. Engaging with people is at the heart of what the job is about and doing so in the College environment is an extraordinary privilege and (almost always!) a great pleasure.

I am grateful for the opportunity to commend this memoir to you.

Sir David Wallace
Master
Churchill College
March 2011

CONTENTS

A BRIEF BIOGRAPHY

Hermann Bondi was born in Vienna on 1 November 1919. Although his parents were Jewish, they were not believers. His mother in particular had a strong dislike of orthodoxy. In his autobiography he describes how these views had to be hidden from the family:

> To keep relations with the rest of the family going one had of course to engage in a certain amount of pretence. Thus we were mildly kosher in our eating at home, but in no way when we were away on holiday. This enabled us to invite other members of the family to come to us for meals, though they might be occasionally a little mistrustful of the strictness of our cooking arrangements.[3]

His dislike for religion clearly originated from these early experiences, as did his ability to interact with people of all religious viewpoints and from all walks of life.

Hermann spent his formative years in Vienna. His interest in walking, skiing and mountain climbing began to develop during these early years, along with his knowledge of the topography of the Alps and his interest in the engineering of railways. He was fortunate in receiving a good education in Vienna. He loved reading, particularly the Greek myths and showed an early aptitude for mathematics, teaching himself calculus at the age of twelve. At school he was soon far ahead of the others in his class.

> Though there was the normal opposition to swots in class, the patent effortlessness of my mathematical successes meant that this caused no hostility at all. I also became rather competent at the spoken word and at hiding from teachers any ignorance that arose from my unwillingness to spend too much time on work that did not interest me all that much[4].

He soon realised that he had an ability to explain complex matters simply and to be good at thinking on his feet, traits that were to become so much a part of his success in later years.

But the rise of the Nazi Party, growing anti-Semitism and the increasingly oppressive atmosphere of late 1920s and 1930s Vienna were kindling in Hermann a desire to live elsewhere. He already considered Austria to be a scientific backwater and knew that an academic career in his home country would be difficult if not impossible, not the least due to the fact that he was from a Jewish family. So he turned his attention to England and the scientific standing of Cambridge and eventually secured a place at Trinity College as a 'Commoner' in December 1936. Fortunately he had both the practical and financial support of his parents and on 19 September 1937 he crossed the Channel to England where, as he says in his autobiography, 'I have lived happily ever after'.

In the months that followed, the political situation in Austria continued to deteriorate and a visit to Vienna over the Christmas vacation 1937 only strengthened Hermann's concern. When on 9 March 1938 the announcement came of a referendum in Austria on whether independence was preferred over union with Germany, he was decided. He wrote:

> … the outcome was clear to me: Hitler could not possibly risk a, for him, negative outcome and would have to intervene by force to prevent this referendum. I therefore sent a telegram to my family of which I am still proud, that they must and without further thought (absolut unbedingt) leave Austria.[5]

Although he was only 18 years of age at the time, his family took his plea seriously. They took the morning train to Budapest, while Hitler marched into Austria the same afternoon. He was always proud that his was one of the few Viennese families never to have lived under Hitler.

Hermann soon settled into life in Cambridge but initially encountered some difficulty, not the least due to a misunderstanding of the Cambridge system and in particular the terms used in mathematics. His confidence was briefly dented when he opted for analysis:

> …thinking that this meant analytical geometry at which I was quite good, whereas in fact it meant the logical foundations of handling functions, of which I was totally ignorant.

Following some hard work, his confidence soon returned and when he presented himself for a second time to his supervisor, the mathematician A S Besicovitch, he knew his earlier confidence had been justified when the latter said to him 'Since you know all this stuff, I will tell you instead of my experiences in the Russian Revolution.'

He went on to graduate with a first class degree and soon determined on an academic career in Cambridge. However, the rise of his academic success was brought to a sudden halt when on 12 May 1940 he was interned as an enemy alien. For someone with such strong anti-Hitler views, this must have seemed a cruel twist of fate. The fact the he never showed any signs of resentment is another measure of his ability to live and work with people.

However, life in internment was in fact to prove a fortuitous turning point. On the first night, in the army barracks at Bury St Edmunds, he met a fellow countryman, Tommy Gold. According to Hermann, Tommy was a bit of a playboy:

> Though only six months younger than myself, he was two years behind me and read engineering, also at Trinity. His intelligence, positive outlook on the world and wide technical interests were immediately apparent. The depth of his thinking and his thoroughness took me a little longer to discover.[6]

They soon became firm friends.

Hermann was later shipped to the Isle of Man and then to Quebec, prior to being released in 1941. He immediately set to work on his PhD on wave theory, but in the spring of 1942 decided to join the Admiralty Signals Establishment to work on radar. However, about this time he heard of a freelance 'think tank' working for the Admiralty headed by a 'wild Cambridge mathematician' called Fred Hoyle and soon afterwards joined his group. Shortly after doing so he managed to persuade Fred Hoyle of Tommy Gold's worth and in October 1942 he, too, joined the group.

The *enfants terribles* of cosmology: Thomas Gold, Hermann Bondi and Fred Hoyle

In 1943, Hermann and Tommy Gold rented a house in Dunsfold, near Witley in Surrey, with Fred Hoyle joining them during the week. Of this period, he said:

> I acquired my first taste there for domestic engineering and for cooking, under Tommy's guidance, but he and I and Fred, when he was with us, spent all our time discussing scientific questions. Fred's enormously stimulating mind, his deep physical intuition, his knowledge of the most interesting problems in astronomy, all combined to give me an outstanding scientific education in the few hours left after a hard day's work.[7]

Fred Hoyle then decided that the summit of Mount Snowdon would be a good base for carrying out trials on radar propagation over the sea and wave clutter and Hermann was charged with running the station. The latter's love of mountains meant that he was to take on this particular role with more than the usual energy and enthusiasm.

His work at the Admiralty Signals Establishment was to continue for three years, although this did not prevent him from writing a Fellowship dissertation in his spare time, resulting in his election to a Research Fellowship at Trinity College in 1943. He said that in those days it was apparently '*infra dig*' to read for a PhD while one was already a Junior Research Fellow', so he withdrew his registration for this qualification.[8]

Following release from war duties in 1945, Hermann returned to Cambridge to the news that he had been appointed an Assistant Lecturer by the Faculty of Mathematics. He was able to return earlier than many since he had been invited to lecture to Newnham College students during the Long Vacation of 1945 (lectures that were also attended by a young Christine Stockman). He was to hold a University Lectureship from 1945-54 and became a naturalised Briton in 1946. In 1947 his path crossed with Christine Stockman again when he came to know her as one of Fred Hoyle's research students. They married on 1 November that year in a civil ceremony in the Shire Hall, Cambridge. When Christine became pregnant in 1948, they moved to a house just outside the centre of Cambridge. Parenthood was a new experience for Hermann and spurred on by the experiences with Alison, their first child, who was born in 1949, they went on to have four more children: Jonathan, Elizabeth, David and Deborah. Mark Goldie describes how Hermann reconciled his work with the demands of fatherhood:

> At home, his working papers were scattered over the floor, where he liked to work sitting cross-legged. His daughter Alison was beginning to walk and was creating havoc. So Alison was put into a playpen. But she screamed mercilessly. The solution was simple: Herman with his papers replaced Alison in the playpen.[9]

But it was the fortuitous meeting with Tommy Gold and Fred Hoyle that was to lead to their greatest contribution to science, the Steady State Theory. While carrying out the work on radar at the Admiralty Signals Establishment, Hermann, Tommy Gold and Fred Hoyle had started to discuss mathematics and cosmology. Of this period, Hermann wrote:

> At the end of the war all three of us returned one by one to Cambridge and continued our habit of discussing scientific questions from morning to night. Naturally the structure of the universe and the oppressive time scale problem occupied us a great deal. Moreover, my own astronomical researches had led to my being asked by the Council of the Royal Astronomical Society to write a Council Review of the state of cosmology. Thus I had to immerse myself in the literature of the day ...[10]

In 1948, the Steady State Theory of Cosmology was published by Hermann and Tommy Gold while Fred Hoyle published separately and almost simultaneously. Hermann wrote:

> Rightly, it is not easy to publish a paper that contains just an idea. Thus we were at a loss how to proceed with an outlook on cosmology we all shared. At this point, Fred Hoyle chose a different path from that preferred by Tommy Gold and myself ... Both Hoyle's paper and our paper appeared at the same time (late 1948) in the Monthly Notices of the Royal Astronomical Society.[11]

The theory was to earn the three of them the nickname, the *enfants terribles* of cosmology. It was also to establish their international reputation, although the hypothesis was eventually displaced in the mid-1960s by the Big Bang theory.

In 1953 the opportunity arose of a chair at King's College, London. Herman wrote:

> My modest experience of the administration in the University of Cambridge, both as a mere customer and on the Faculty Board, left me with a thorough distaste of the whole matter. This was so strong that when the offer of a chair at King's College, London, came, at a time when I had a very comfortable tenured position in Cambridge, an important point in my acceptance was that I would not be Head of Department and thus be free of all administrative duties. My move to London was also prompted by my fear of getting stuck in Cambridge, by our desire to live rurally, as the London commuter system allows one to do and not least by the friendly robust charm of the then Principal of King's, Peter Noble.[12]

He took up his new post in October 1954, purchased a house in Reigate, which he and Christine subsequently renovated, and did not leave until taking up the Mastership of Churchill College in 1983. The move to King's College, London, was a great success but was also to prove yet another turning point in his career. Notwithstanding the promise of no administrative duties, Hermann soon found himself chairing meetings and discovered that he was not only good at it, he enjoyed it as well. During this period he was also invited by the Secretary of State for Air to sit on the Meteorological Research Committee. Having no connection with meteorology but understanding his role was that of an outsider, he accepted. Of this experience he says:

> My original academic inclination to think that I could contribute only in something of which I had real mastery was finally shattered.[13]

Other offers followed and Hermann soon found himself very much involved in government science as well as in his own teaching and research. Mark Goldie writes that the 1950s and 1960s were in fact to see his most productive scientific period:

> … with a stream of papers on gravitation, relativity, waves, interstellar matter and astrophysics. He in fact regarded his papers on gravitation as his more important. Cosmology is now populated by Bondi principles: the Bondi mass, Bondi waves, Bondi accretion and the Tolman-Bondi universe.[14]

But he was also expanding his horizons elsewhere. He was appointed secretary of the Royal Astronomical Society from 1956 to 1964 and was awarded the Royal Astronomical Society gold medal in 2001. During the 1950s Hermann and Christine both joined the British Humanist Association, with Hermann being appointed President from 1982-99. There was also a request to report on the proposed Thames Barrier to protect London from flooding. This particular report was completed in 1967, the barrier finally being opened on 8 May 1984 and he was always to regard this as one of his major achievements.

In July 1967 a call came from the European Space Research Organisation (ESRO) to see if he would be interested in the position of Director-General. This was another challenge that was of interest to him and one that he was again willing to meet:

> This was of course an entirely new challenge for me and I must confess I did not realize how big a challenge it was going to be. After what I had done I was full of self-confidence, but there were many moments in ESRO when I felt stretched to breaking point. This was responsibility of a far greater order than I had ever experienced before, although when I had said "Yes", I had not totally appreciated this. I think I can honestly say that I had never enjoyed a job more than ESRO, particularly perhaps because it was so challenging.[15]

As his three year appointment was drawing to a close, Hermann was approached by the newly appointed Secretary for Defence, Lord Carrington, to see if he would be interested in taking over as Chief Scientific Adviser to the Ministry of Defence, a post he was to take up on 1 March 1971. Hermann wrote:

> The challenge in Defence was wonderful. The system was superb in making me and my small group independent of the chief procurement agencies, the Government Research Establishments and the operational requirement staffs of the individual forces. So when projects had reached the stage where a Yes or No was required for their development, before large funds could be spent, they had to come before the Defence Equipment Policy Committee which I chaired and greatly influenced. In this position

therefore I was in no way tied by previous discussions. I and my staff could look at the issues in an entirely independent manner and come to our own technical conclusions, ask our own searching questions.[16]

His efforts did not go unnoticed and in 1973 he was awarded a KCB (Knight Commander of Bath). Of this recognition of his achievement he wrote with his usual modesty:

Though at first I was not too sure about my attitude to it, this was soon settled when I saw the pleasure it gave to others and the added weight it gave to my actions.[17]

Notwithstanding his usual heavy workload, it was also around this time that he became involved with the Institute of Mathematics and its Applications, taking on the role of President in 1974 and 1975. His successor as President was Dame Kathleen Ollerenshaw. He wrote:

It may be worth mentioning here that somewhat later she stimulated me to work with her on a mathematical subject far away from my usual fields, namely Magic Squares and we published jointly a major paper on it in the Philosophical Transactions of the Royal Society in 1982. She has always been a person of immense energy, wholly undiminished by age.[18]

Sir Hermann remained in defence until 1977 when he was asked by Tony Benn to become Chief Scientist to the Department of Energy. Ian Roxburgh writes of this period:

Much of his contribution here consisted of advising the Secretary of State on research programmes of the nationalised industries: electricity, coal and gas; and research programmes in renewable energy sources, especially wave energy. But two jobs were the most rewarding: leading the UK delegation to the International Fuel Cycle Evaluation (INFCE) and the Severn Barrage Committee.[19]

While Sir Hermann was to write:

Hardly had I joined the Department when the issue of the International Fuel Cycle Evaluation arose, which at President Carter's initiative got the world thinking about how to permit the growth of nuclear power while preventing the proliferation of nuclear weapons. I was asked to take the lead for the UK and for 2½ years was again deep in diplomacy which I enjoy. The effort was a considerable success in avoiding a series of real dangers, such as a split between the USA and her allies and a worsening of North-South relations.[20]

Ian Roxburgh notes that he was then appointed to head the Severn Barrage Committee:

It is indicative of Hermann's chairmanship skills that this very heterogeneous committee, with representatives of water authorities, ecologists and parliamentarians reached a unanimous report which recommended a 16km barrage be built.[21]

The Bondi Committee report, as it became known, was published in 1981. Although its recommendations were not acted upon, the plans were strongly built on a few years later by the Severn Tidal Power Group.

In September 1980, however, Sir Hermann experienced his first retirement. Under Civil Service retirement rules he had to leave his post, but was fortunate in being invited to take on the new role of Chairman and Chief Executive of the Natural Environment Research Council (NERC). Of this, he says:

… an organisation small enough (some 3000 staff) for me to feel the full load of personal responsibility and involving a whole host of sciences utterly new to me such as geology, oceanography, terrestrial ecology and marine biology. Once again I have the delight of learning.[22]

This was another job he was to enjoy immensely. Mark Goldie notes:

… he also knew how to stand up to political pressure: his refusal to follow Prime Minister Margaret Thatcher's policy injunctions for NERC probably cost him a seat in the House of Lords.[23]

His term of office at NERC drawing to a close in 1984, just before his sixty-fifth birthday, posed a problem to him since for ordinary staff the age limit was sixty. To seek to prolong his own appointment there in the face of such a policy was to him a cause of some embarrassment and a course of action he was not inclined to follow. His second retirement was therefore approaching. However, in the spring of 1982 he was asked if he would be interested in his name being put forward for the Mastership of Churchill College. This was a new challenge and after some deliberation, it was one he decided that he would be willing to take.

In Churchill College, the College Officers were busy making the arrangements for the election and arrival of a new Master. So it is to Churchill College we now turn, to the next chapters in Sir Hermann's life and career.

Christine and Hermann Bondi on their wedding day (1947)

Sir Hermann Bondi (c. 1985)

THE THIRD MASTER

So far, I think, I have managed to avoid getting into a rut, something I never wanted to do, for a rut is just like a grave, only longer.[24]

As Registrar I greatly enjoy the challenges of my role. I take pride in what I do and strive to achieve a job well done and if, after twenty-three years, I have driven a furrow, it is a furrow that suits, but never a rut. I am fortunate, too, that if asked which of my jobs I have most enjoyed, I can say that it is working at Churchill College.

I first came to Churchill on a late summer's day in 1988 as the 'temp' from the agency. Sir Hermann needed a new secretary and the agency sent me. Having worked for two Masters of Emmanuel College before my sons were born, I was no stranger to the role and soon got to know his way of working. I liked him right from the start and was charmed by his jokes and stories. The permanent position being vacant, I applied and was appointed to the role. So it was that Sir Hermann became my third Master.

At the time of my arrival in 1988 the process for the selection of Sir Hermann's successor was well under way. Files in the College Archives[25] reveal how the election process has evolved over the years, while other files depict the domestic activities that were taking place. These, together with details from Sir Hermann's personal files, provide a snapshot of the activities within Churchill College both prior to and shortly after Sir Hermann's appointment as third Master of Churchill College and it is to these I now turn as the starting point of my story.

Along with Trinity College, Cambridge and Christ Church, Oxford, the Mastership of Churchill College is a Crown appointment. It is an appointment made by the Monarch on the advice of the Prime Minister and is a matter determined in collaboration with the Appointments Secretary at Downing Street.

The College has an elaborate process for selecting nominations for the Mastership, with the burden of responsibility falling on the Vice-Master. The body which votes for the Master is technically 'the Fellowship' and not the 'Governing Body', since officially it is only advisory (making nominations) and not a decision of the Governing Body. A Nominating Committee has to be set up to seek and consider proposals from Fellows and at least three formal meetings of the Fellowship will need to be convened. Numerous meetings will need to be held with small groups of Fellows to discuss possible candidates and a series of long lists and shortlists will also need to be drawn up. Mary Beveridge was College Secretary at the time of the election to the third Mastership and she attended all the meetings of the Nominating Committee with the Vice-Master, Jack Miller, as well as dealing with all the paperwork and note-taking. She recalls that the first meeting was held on 4 February 1982:

> We worked out a procedure – which has been used for every changeover of Master since then – for the Fellowship to consider possible candidates and for these names to be put forward to the Appointments Secretary. The names would then in due course come back to the College from the Prime Minister's office, together with the names of any official candidates put forward by the PM, to be

The ballot box: still in use at Governing Body meetings today

voted on by the Fellows. The recommendation of the Fellowship would then be submitted to Number 10 for – it was hoped – Prime Ministerial approval. We were fortunate in 1982 that Argentina invaded the Falklands, leaving Mrs Thatcher with little time to spend thinking of possible candidates for the Mastership of Churchill College.

The process culminates in a formal vote by the Fellowship. The records in the College Archives reveal that a number of voting methods and formulae had been considered by a group of Fellows two years previously and that of the single transferable vote was agreed to be fit for purpose. A document dated 3 November 1981 outlines the various methods under consideration, as well as the degree of attention paid towards this single aspect of the procedure:

> Let c be the number of candidates and let q be the minimum number of Fellows that must support a recommended candidate under paragraph 5 of the general procedure. Let d be $n-q+1$, where n is the number of Fellows present...[26]

Looking through the files to see if the same procedure was used in subsequent elections, a seven hundred word explanatory document issued to Fellows in 1995 indicates that some modifications have been made:

> Let q be the minimum number of Fellows that must support a recommended candidate. Let d be $n-q+1$, where n is the number of Fellows voting. This is the smallest number of Fellows that can ensure that a candidate is not recommended, by leaving that candidate unmarked on their ballot paper. During the counting procedure, one or more candidates may be deleted from the count. From then on, each ballot paper shall be treated as if the remaining candidates had been the only candidates. In particular, the boxes on each ballot paper shall be re-numbered in the same order but with consecutive numbers from 1 upwards.[27]

Mary Beveridge recalls:

> The very complicated ballot system was devised by Drs Dixon and Tristram. The College carpenters designed two special ballot boxes into which the Fellows would cast their votes – one for postal votes (since it had been agreed that Fellows could vote from whichever part of the world they happened to be in when the time came) and the second to be used by Fellows voting in person. Both boxes were kept under strict security in the Bursar's office!

As matters progressed it would have fallen on the Vice-Master, Jack Miller, to write to potential candidates to see if they would be interested in taking on such a role and this he duly did. Christine Bondi says that while being delighted at being contacted by the College, Sir Hermann was initially not at all certain that the Mastership of a College was something he wished to do. In his autobiography he was later to write:

> In Spring 1982 I got a letter from Churchill College, Cambridge, asking whether I would be interested in my name being put forward for its mastership. This was by no means the first such invitation I had had, but I had rejected all the others readily. First, I thoroughly enjoyed having a full-time, demanding managerial task, whereas the headship of a College is certainly not full-time. Secondly, great though the honour is, my memories of the overall administrative atmosphere of Cambridge were not over happy. Third, we were rather committed to the maintained school system and Cambridge was to a large extent populated by products of independent schools. Yet the invitation from Churchill looked different to my wife and myself.[28]

Sir Hermann goes on to say:

> Our visit to Churchill on the early May Bank Holiday of 1982 totally charmed us and won us over. Churchill is a college modern in outlook as well as structure, a college full of friendliness, a college where the large majority of the undergraduates come from the maintained sector, a college with strong international links. What more could we ask for?[29]

The interest of all candidates being confirmed, the College was now ready to make its choice. The vote got under way. Mary writes:

> My 1982 diary notes that Sir Hermann Bondi came top of the ballot which was run on 3 May 1982 and the Nominating Committee met on 7 May to decide on the procedure to be followed thereafter. (The official ballot closed at 5.00 p.m. on 6 May.) The result of the ballot was reported to the Governing Body on 14 May and handed over to Robin Catford, the Appointments Secretary, on 18 May.

On 14 June 1982, Sir Hermann received a letter from the Prime Minister, Margaret Thatcher, informing him that she intended to recommend to The Queen the appointment of a successor. The Prime Minister wished to know if he would be willing for her to submit his name to Her Majesty and by way of encouragement wrote:

> I am sure that your appointment would be widely welcomed and that you would continue to develop the traditions of the College in the way that Sir Winston himself would have wished.[30]

This letter was clearly treasured by Sir Hermann as it remained with his private papers and was not transferred to the Archives Centre on his retirement as Master. Furthermore, by now he had had sufficient time to consider the practicalities of such a move and he wrote to the Vice-Master to discuss this further. Jack Miller responded on 22 June 1982:

… the fact that your present appointment with NERC [Natural Environment Research Council] does not end until 30 September 1984 and that you would have to give priority to those duties, was known to the Governing Body before your name was put forward and I am sure that there should be no difficulty in making arrangements acceptable to both parties.[31]

Following this reassurance, Sir Hermann sent a hand-written letter to the Prime Minster on 12 July 1982:

Dear Prime Minister,

Thank you for your letter of 14 June about the Mastership of Churchill College. I shall be very grateful if you will submit my name to her Majesty.

There are just two points I would like to make: First, I definitely wish to continue as Chairman of NERC to the end of my 4 year term (30 Sept 1984). Thus there will be a year of overlap. The College have made it very clear that they would much prefer a Master who is absent for much of the time to postponing the start of my Mastership. During this period of one year (during which I will of course not receive any salary from the College) I would plan on spending numerous weekends in the College and thus get to know both staff and students. I would also expect that my NERC duties would permit me to chair a number of the College's Executive Council meetings.

My second point is that I would wish to be consulted before a public announcement is made so that I can inform NERC staff in my own way.

Yours sincerely
Hermann Bondi[32]

It is clear that Sir Hermann's sense of responsibility towards the staff of NERC (Natural Environment Research Council) was central to his deliberations at the time and in the files there are numerous handwritten drafts and re-drafts of a suitable announcement to be issued to his staff. He was greatly concerned that the confidence and morale of NERC staff should not be undermined by the fact that he would be taking on the Mastership of Churchill College during the final year of his Chairmanship of NERC and he made strenuous efforts to ensure that both NERC and the College were in no doubt as to his total commitment.

Shortly after the announcement was made to NERC staff and his appointment became public news, Sir Hermann received a flood of letters of congratulation.[33] The letters exemplify the disparate and universal nature of his work and interests at this time, many coming from within Cambridge or from those with close connections to the College.

On 31 August 1982 Sir Hermann received a letter from the Chief Scientist and Deputy Secretary, Department of the Environment and Department of Transport:

I am sure that you will find the mastership of a college thoroughly agreeable and it will be good for all of us to have you around Cambridge. I am relieved to see from the papers that you will be carrying on at NERC until the completion of your period of office and only hope that the strain of combining the two important tasks does not erode even your vitality! Perhaps we can do something to improve Cambridge – Swindon rapid transport systems.

While on 15 November 1984, B I Edelson, Associate Administrator for Space Science and Applications, NASA, Washington DC wrote:

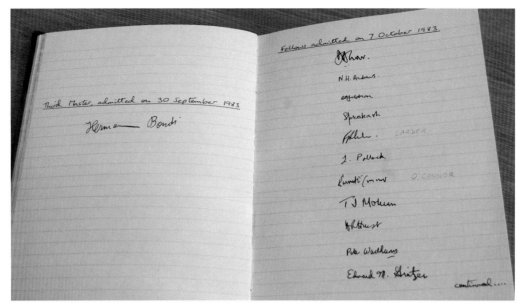

The Admissions Book which contains the signatures of current and former Masters and Fellows of the College. It is still in use today

I have noted your most recent move back into academia with interest. You certainly have had the most fascinating positions and your new one seems exciting and challenging. I visited Churchill College about 18 years ago and was very impressed by its modern architecture and its philosophical approach to education. I'm sure you'll lead its march toward even greater goals.

Sir Alan Hodgkin, Master of Sir Hermann's old College, Trinity, felt that his election as Master was '… excellent news - good for Churchill and Cambridge – and nice to have another Trinity man in our galaxy as heads of houses ….', while a card from 'Ian' dated 3 September 1982 expounded:

Hail, mightier than Hodgkin!
For though he is Master of Trinity, you will be Master of Churchill,
Le Frère Aîné de la Trinité.

On 21 October, one of the College's former Honorary Fellows, Pyotr Kapitza, wrote to Sir Hermann as follows:

For me the fate of this College is very close. The first Master of Churchill College Sir John Cockcroft was a great friend of mine. The progressive and advanced method with which Cockcroft organized the College was very familiar to me and I always supported his ideas.

I hope you will develop the fates of the College in the same line and I wish you best opportunity.

In a letter dated 23 August 1982, another former Honorary Fellow, Lord Todd, gave some advice based on his own knowledge of the College:

I have just had a note from Downing Street about your appointment to Churchill and I send my congratulations and best wishes. I believe you will find it both a rewarding and enjoyable task – I certainly did at Christ's and Churchill is a very friendly outfit. I don't expect there will be much need for help, but please remember that I am at your disposal. I look forward to your coming and I shall now try to discharge my responsibilities as an Honorary Fellow with more regularity than I have in the past!

A former Vice-Master, Jack Pole, wrote on 7 September 1982:

> May I take the liberty of writing to express my pleasure at your appointment to be Master of Churchill? My qualifications for the intrusion are that I held a fellowship there for sixteen years (1963-1979) and ended up as Vice-Master of the College. I go back periodically and hope I shall one day have the pleasure of meeting you there. Churchill is not only high-powered intellectually but is a fascinating and very congenial community, though in some significant ways I feel that it is still in process of formation. As one who is fond of the College I still felt when I left that it hadn't yet clearly resolved some of the problems of its own identity. I believe that to a greater extent than is usually the case when taking up the headship of a college, you will be in a position to make a contribution to that and I am sure that everyone who cares about the College will be delighted.

Back in Churchill College discussions had been taking place for over a year with respect to the arrangements for Sir Hermann's admission as Master and his move to Cambridge. On 23 September 1982 a draft Governing Body paper enclosed with a letter to Sir Hermann from the Bursar, Hywel George, outlined a number of issues that the Governing Body would need to consider in view of the fact that Sir Hermann would be continuing as Chairman of NERC until September 1984.[34] The paper noted in particular that 'during the first year of office the Master will continue to reside in Reigate but he and Lady Bondi will spend a significant number of weekends in Full Term at the Master's Lodge'.

Hywel George wrote to Sir Hermann on 8 February 1983:

> We are considering what changes, if any, have to be made to our administrative arrangements next year and in particular, how to staff the Master's office…

> …Under present arrangements, the Master's secretary deals with the Master's office, his fairly light social programme and his extensive non-College activities. Some of the Master's College work, especially that concerned with the College Council and its committees and the Fellowship Electors, is dealt with by the College Secretary, Miss Beveridge, who with the help of a part-time secretary looks after my office and that of the Vice-Master, as well as handling the work of the main College bodies and committees.

> How we operate after the first year depends very much on your activities and requirements, but it seems to me that until October 1984, what we need is someone to:

> (a) deal with the routine work of the Master's office, bearing in mind that for the first year you will not be able to be involved in routine administration;

> (b) arrange a social programme for your visits to the College;

> (c) prepare briefs on important College issues before your visits to the College.

> As College Secretary, Miss Beveridge will be able to help with preparing briefing papers on College affairs, but other arrangements will have to be made to deal with the entertainment and general office work …[35]

Toasting the new Master (1983)

Sir Hermann responded to Hywel George on 17 February:

> I think that you have admirably summarized what I would require from a secretary during the first year of my appointment. For this period my Private Office in NERC will continue to deal with my overall timetable, co-ordinating arrangements for both my non-NERC and my NERC engagements.

> After this first year it is much more difficult to forecast the nature of my requirements. I would expect that a large number of demands on my time will be made from outside College; even if each of these was to be modest, it would not necessarily be entirely easy to arrange my diary. I would expect my secretary to handle both liaison and co-ordination of my diary. However, if any of these non Churchill engagements should make a major demand on my time, I would expect the organisers of it to furnish some secretarial assistance.[36]

Sir Hermann Bondi was formally admitted to the Mastership at 6.30 p.m. on Friday 30 September 1983. In accordance with Statute XIII (Appointment and Duties of the Master), Sir Hermann was admitted by the Vice-Master, Dr Jack Miller and the Fellows after having read aloud in their presence the following declaration:

> I, Hermann Bondi, appointed Master of Churchill College, do solemnly declare that I will endeavour to the utmost of my power without fear or favour, to promote the interests of the College as a place of education, learning and research.[37]

Sir Hermann then signed the Admissions Book, which is still in use at Governing Body meetings today and so became the third Master of Churchill College.

CHURCHILL COLLEGE

THE ONE HUNDRED AND EIGHTY-SIXTH MEETING OF THE GOVERNING BODY
will be held in the Fellows' Dining Room on Friday 14th January 1994 at 5.30 p.m.

AGENDA

1. Apologies for Absence

2. Admissions

   Professor G Steigman

3. Confirm Minutes: 185th Meeting (Statutory) of 3rd December 1993 (previously circulated)

4. Matters to Report

5. College Council Election for 1994 and 1995

   There is one vacancy for 1994 and 1995. Two nominations have been received:

   Dr B L Gupta   Proposed: Dr Allan   Seconded: Dr Hoskin

   Mr A J Pauza   Proposed: Dr Milne   Seconded: Mr Allen

6. Report on the Operation of the Møller Centre for the Michaelmas Quarter

7. Loans to the Møller Centre

8. Special General Meeting of the Møller Centre for Continuing Education Limited

   (a) Approval of the notice given for this meeting

   (b) Approval of the increase in authorised share capital to £2 million £1 shares

9. Appointment of Inspectors of Accounts

10. Council Business

11. S.C.R. Matters

    To be raised by the President

12. Any other Business of which prior notice has been given

    Any Fellow who wishes to raise any matters under this item is asked to notify the Bursar no later than 12th January 1994

13. Date of Next Meeting

    Friday 11th March 1994

Please send Apologies if you are unable to attend

ANY OTHER BUSINESS

Soon I learned that the task of diplomacy is not to be secretive, but to be clear, a task for which university teaching is an excellent preparation.[38]

Not including the JCR and MCR (student) committees, there are currently in excess of forty committees operating across the College, all with their own terms of reference, formal memberships and particular *modus operandi*. Despite attempts across the decades to reduce the number of committees in the College, the number of committees in the 1980s would probably not have been dissimilar to the number in operation today. While the greatest burden of committee attendance in the College rests with the Bursar, the Master is still required to chair a number of internal committees. Indeed, Statute III.3 clarifies the duties of the Master as follows:

> The Master shall exercise a general superintendence over the affairs of the College and shall preside when present at all meetings of the Governing Body and of the Council and at all meetings of other bodies or committees of which he or she is a member and shall, except as otherwise provided in these Statutes, be entitled, in the case of an equality of votes, to give a second or casting vote. The Master shall have the power, in all cases not provided for by these Statutes or by order of the Governing Body or the Council, to make such provision for the good government of the College as he or she shall think fit.[39]

Sir Hermann was certainly no stranger to committee work. According to an article published in 1981, when he left academic life in 1967 his secretary had to send his resignation to 84 committees,[40] and he made clear from the outset that he wanted to attend as many meetings and events at Churchill during his first year as was humanly possible. But without the benefit of electronic diaries and quick and easy communication by email, it was inevitable that this was always going to be a real challenge. It was also a challenge that would not be achieved without significant effort on the part of Sir Hermann's Private Secretary at NERC, Jonathan Bates, and the Bursar, Hywel George, at Churchill. In a letter to Hywel George dated 25 March 1983, Jonathan Bates listed in some detail the various meetings that Sir Hermann would and would not be able to attend. He added:

> Sir Hermann asked me to emphasise to you that with all these dates he remains rather at the whim of Ministers, etc., as long as he is a Government employee – if he was asked to see the Secretary of State at short notice he would, of course, have to give this some priority. I am sure you will understand this. Nevertheless, Sir Hermann intends to do his utmost to be at Churchill on the dates indicated above and on others.

> Please do not hesitate to contact me at any time to discuss Sir Hermann's timetable or his movements. It may well be that there are occasions on which we can meld the requirements of NERC and of Churchill in such a way as to be of assistance to you. In addition, Sir Hermann has a most capable team here at Swindon and we would all be glad to be of help to you during this slightly difficult one year period while Sir Hermann occupies two Chairs.[41]

Top space scientist to head Churchill

Space-age mathematician Prof Sir Hermann Bondi has been appointed the new Master of Churchill College, Cambridge.

He will take up the appointment in July 1983 when the present Master, Prof Sir William Hawthorne retires.

Sir Hermann, aged 62, is Professor of Mathematics at King's College, London, and the first full-time chairman of the Natural Environment Research Council.

Sir Hermann moulded European space research and the ESRO later developed into the European Space Agency as it is known today.

Policy

He joined the MoD in Whitehall at a time when a new framework of Government-funded research and development was being introduced and the move to the Department of Energy involved formulating the research policy on issues such as the exploitation of the North Sea, and coal and nuclear energy.

He was born in Vienna and graduated at Trinity College, Cambridge. He was a lecturer at Cambridge University and in 1967 to 1971 was Director General of the European Space Research Organisation.

Sir Hermann became chief scientific adviser at the Ministry of Defence in 1971 and in 1977 moved to the Department of Energy where he was chief scientist until 1980.

As the chairman of NERC, he heads the organisation responsible for a range of research groups, including the Cambridge-based British Antarctic Survey and the Monks Wood Institute of Terrestrial Ecology near Huntingdon.

Sir Hermann, who has published a number of works on Cosmology and astrophysics, at present lives in Surrey.

Sir William, 70, an eminent scientist and technologist, has been the Master of Churchill since 1968. He was a pioneer of the jet engine and recently won a Royal Medal for his work on aerodynamics.

Sir Hermann's appointment attracted the interest of the media (1982)

These difficulties had not come as any surprise to the College, of course. In a draft Governing Body paper dated 23 September 1982, Hywel George had noted the following:

During the first year of office the Master expects to attend all meetings of the Governing Body, about 60% of College Council meetings and the more important meetings of the Fellowship Electors. He will receive papers on College business and will give his comments and views by letter or by telephone. He will not be able to attend meetings of other bodies and committees of which the Master is normally a member.

He expects to attend the main College social functions, i.e. the Dinners for Freshers and Advanced Students, College Guest Nights and the Graduands Garden party. During their visits to the Lodge at weekends, the Master and Lady Bondi hope to meet and entertain the Fellows and as many Junior Members and staff as possible.[42]

Such was his conscientiousness and commitment to his new role and his concern about the certainty of his presence, that in the first sentence of each of the first two paragraphs above, Sir Hermann had crossed out the word 'expects' and written 'intends' in its place.

Statute VI (Absence of the Master) requires that during the absence of the Master, the functions of the Mastership be performed by the Vice-Master and Jack Miller worked tirelessly during 1983/4 to ensure a smooth interregnum. He wrote:

Prior to Hermann's arrival, there was a short interregnum which left me acting as Master. With a lot of help from members of the Council and from the Bursar Hywel George in particular, I just about survived.

Of this interim period, the Bursar, Hywel George, recalls:

It was left to the Vice-Master, with the help of the College Officers, to carry out most of the duties of the Master and to ensure the smooth running of College affairs. This placed an additional burden on him but his workload was eased by the fact that he had already served four years as Vice-Master, his views on most College matters were well known and it was possible to draft papers and letters for his signature without further reference.

The College Secretary, Mary Beveridge, agrees:

Professor Kapitza in Cambridge ?

3.45 pm   Mr J Mobed
4:30 pm   Mr N R Williams

[4.45 pm? Eaden Lilley booked for Fellows' photographs]

5.30 pm   Governing Body
     —   meeting with Professor Kapitza
7.30 pm   High Table and Informal Dessert for New Fellows

A diary extract in Paula Halson's handwriting

Jack used to appear regularly in the office at lunch time asking "What's new?" and would be brought up to date by the Bursar or me on any particular bits of College gossip or business. I well remember him signing many sheets of College headed paper so that letters could be typed later to be sent out over his signature! We got quite good at guessing whereabouts on the page his signature should be placed, depending on the probable length of the letter (right down at the bottom for lengthy letters of appointment for new Fellows, somewhere in the middle for letters of congratulation …).

During 1983/4 another new pattern of work was to evolve. Mary notes:

If he [Sir Hermann] was here on a Friday for a Governing Body meeting he would normally stay overnight in the Master's Lodge and deal with any College business on the Saturday. I used to provide him with advance briefings for any meetings he attended – an arrangement which suited him well as he was used to having a Civil Service type private office at NERC and which continued throughout his Mastership.

On the matter of support, Hywel George expands further:

During his years in public service he had been served by a private office of several people and he was used to a high level of support. Churchill endeavoured to continue this system but on a more modest scale. It was nevertheless possible to supply Sir Hermann with briefing notes and papers on most College issues and to steer him gently in the 'right' direction. With this in place he was always an easy person to work with.

Despite his good intentions, the first Council meeting that Sir Hermann was able to chair as Master was the 595th Meeting of the College Council held on 14 February 1984. To his great regret he was unable to chair another meeting of the College Council until the beginning of the next academic year. However, in 1984/85 he went on to preside over 14 of the 16 Council meetings that took place during that year, a sign that his joint diary commitments were now easing and that he was finally able to focus his attentions fully on the job in hand.

The Agenda and papers for that first Council meeting on 14 February 1984 would certainly have given Sir Hermann a flavour of what was in store, as topics at Council meetings inevitably range between the strategic as well as the more prosaic. At this first meeting, the Senior Tutor reported on the poor behaviour of a Churchill sports team where members were alleged to have damaged another College's property, while the President of the JCR reported that the proposal that the video games machines should be removed from the Buttery Foyer during vacations had been unanimously rejected at the last JCR Open Meeting. Council had also considered the development of a strategy for higher education in the 1990s, as well as the Admissions Tutor's report on undergraduate statistics which indicated that the number of women applying was still low compared with other mixed Colleges. The meeting also considered a proposal by a current Fellow, Andrew Tristram, to form a group of Fellows and students 'who would be interested in using the computer facilities' and it was at that first Council meeting under Sir Hermann's chairmanship, that the College's Computer Committee was established.[43]

Of those who responded to my request for anecdotes and recollections of Sir Hermann for this memoir by far the majority I received related to his style in meetings. They all reveal that he had his own particular *modus operandi*. Jack Miller's widow, Marcia, recalls that Jack felt that he learned a lot about running meetings from Sir Hermann and greatly admired his style. Jack himself had written:

> I saw then how meetings should really be run. His ability to guide a discussion so that we did not indulge in the usual academic habit of wandering off the topic was outstanding. His summing up of the discussion so that we knew what we had agreed was something to behold.

Hywel George says:

> Sir Hermann's public service experience was reflected in his handling of Governing Body, Council and Committee meetings. He was never obviously forceful but listened patiently to the views of his colleagues often delivered at considerable length. He moulded people gently and gradually, often with the help of an Austrian folk story, until they came to his way of thinking and made sensible decisions. The recording of discussions was helped by his acute summaries at the close of each discussion.

Others noted his expertise in finding solutions and compromises to complex issues. A former Fellow, Mark Tester, recalls:

> … in meetings one very notable characteristic was his desire to always try to find a compromise solution and to avoid the need to vote on issues. This admirable trait was thwarted in one Governing Body meeting over one issue – I cannot remember if it was keeping the College accounts in Barclays (during the South African anti-apartheid protests) or if it was to do with whether to ban The Sun from the SCR – anyway, a compromise solution was simply not possible; it was either one thing or another.

So a vote was held. When the results were handed to Sir Hermann, he laughed uproariously, triumphantly waving the results and saying this was a vindication of his dislike of voting on issues in committee. The vote was a draw. The Governing Body was perfectly split on the issue!

Similarly, Jonathan Parker, who was President of the MCR from 1988/89, recalls:

I remember how skilled he was in dealing with tricky issues on the College Council (e.g. we had an issue related to funding coming to College from a source inside South Africa during the time of apartheid) and drawing conclusions from discussions where I could see none!

By way of contrast, Mary Beveridge suggests that Sir Hermann's summaries were not necessarily in accordance with preceding discussions, although she admits that this practice did make her life easier:

… as it meant that the minutes were virtually written in advance – and then completed by Hermann's habit of summarising a discussion at the end of it. Although it has to be said that Hermann's summaries were not always completely in accordance with the actual discussions which had preceded them – a practice in the tradition of the apocryphal Cabinet Secretary at the end of Cabinet Office meetings:

> The Cabinet rises and goes to its dinner,
> The Secretary stays and gets thinner and thinner –
> Racking his brains to record and report
> What he thinks that they think they ought to have thought …

and which, happily for me, meant that when I came to write up the minutes I was not left racking my brains trying to interpret the thoughts of the Fellowship. Meetings chaired by Hermann were frequently enlivened by the Viennese stories of which he had an inexhaustible supply and which he used to produce at tense moments to lighten the atmosphere.

In terms of his style of chairing meetings, Alison Finch, Fellow and Vice-Master at the time of writing, writes:

He had a delightful way of chairing meetings that was simultaneously business-like and relaxed – I remember half-way through one long meeting he announced, "And now for a technical break," and left for the loo – as did some of the rest of us.

While Michael Allen, Fellow and former Bursar, remembers:

I think my overwhelming memory of Hermann was of his brilliant summarising of debates at Council and GB. He was a past (M)master at appearing to be very busy with algebraic doodles on his note pad, or even appearing to be quietly dozing, when suddenly he would spring into action and dynamically summarise a complex debate in a highly succinct and complete manner. Always a master stroke of which I had a very clear view sitting in the Bursar's seat on his left hand.

But perhaps Sir Hermann's style is best summarised in a letter from Neil Andrews, former Fellow and Dean, dated 6 December 1989:

I thought I would draw a line under five years of membership of College Council by saying how much I have admired the patient and fair Chairmanship you have exhibited during all those years. As a lawyer

I am really a student of such public performance and I feel that I have benefited a great deal.

For my part, as a verbatim minute-taker I would often find myself silently thanking Sir Hermann for these succinct and eloquent summaries, as they would often bring together somewhat complex and conflicting viewpoints. And I can attest to the algebraic doodles on the Agenda to which Michael Allen refers above. When meeting papers were returned to me for filing, I could sometimes calculate the importance or otherwise of the discussions depending on the density of the doodles across the page.

Over the course of his Mastership, Sir Hermann's diary never appeared to ease and Hywel George had to work hard each year to secure dates for College events and meetings. Even towards the end of his tenure, Sir Hermann's diary continued to be exceptionally full. Back in 2005 while clearing out his College room after his death, I found a document which I had typed in 1988 and which listed some of his non-College appointments for the year. For example, the entry for March 1988 was as follows:

1-2 March		Brussels (NNE)
7 March		Master's recording for Leicester Univ.
9 March	pm	Bedfordshire Mathematics – Master to lecture
10 March		Annual Luncheon, Parliamentary & Scientific Committee – Savoy Hotel
14 March	am	History of Radio Astronomy interview
15 March	pm	Lloyds Council Lunch
16 March	am	CEED Seminar at Royal Overseas League, London
	pm	Talk to Ministry of Defence Staff
	pm	Talk to Institute of Marketing
17 March	pm	MOD Interviews
22 March	pm	Address Norwegian Science Group & Minister
24 March		Visit Sir Karl Popper
28 March		Brussels (NNE)
29 March		Brussels
30 March	pm	MOD Interviews, London

Referring back to the College's minute books which record its proceedings across the years, it seems that notwithstanding these appointments, Sir Hermann was able to chair 18 out of 18 Council meetings in 1988, while a note in his handwriting at the foot of this document indicates that between 4 January and 26 June 1988, he also delivered twelve Tripos lectures on Cosmology, attended five meetings on Management Education and attended four meetings as Chairman of the Faculty Board of Education.

Even after his retirement Sir Hermann continued as an active and valuable member of the Estates Committee (the Buildings Committee and the Grounds & Gardens Committee amalgamated in 1992) and remained on the Committee until his death in 2005. In 1981 he had written:

By minimising my effort except on issues where I had something to offer, be it knowledge, experience or just opinions, I could come in on these with verve or, as I prefer to put it, with all four feet.[44]

Jennifer Brook (formerly Rigby), Fellow and current Bursar, can attest to this. She writes:

Hermann performed an extremely useful role on the Estates Committee right up to his last few months. He was our secret weapon. When expert consultants were brought in to advise us on structures, or concrete deterioration, or district heating systems, he would listen carefully, while giving the appearance of dozing quietly at one end of the table. They would say their piece and other members of the committee would question them. Hermann would then announce something along the lines of: "I remember, when we were looking at the forward compartment of the Hunter class of nuclear submarines, there was always a problem of too much heat/too little heat. You should consider........." His anecdotes were often a lot more amusing than this and always relevant and apposite. You could see the consultants suddenly "sit up and take notice", realising themselves to be in the presence of a truly wise man. Moreover, I would get regular calls in the office from Hermann, when he remembered something about the buildings and related problems which he thought I should follow up: he was seldom wrong.

Sir Hermann and Lady Bondi outside Newnham College (2004)

Sir Hermann and Lady Bondi preparing to entertain guests in the Master's Lodge dining room (1986)

LIFE IN THE LODGE

Many a time I would walk into the study and find Sir Hermann working at his desk with Connie, the cat, who would be draped around his neck like a scarf.[45]

Sir Hermann and Lady Bondi were invited to view the Master's Lodge some time before they were to move in. They had some initial concerns about its furnishing and suitability. Sir Hermann wrote to the Bursar, Hywel George, on 17 February 1983:

> … There is the question of the Lodge and of making it not just habitable for us by next October, but suitable for entertaining. The two major questions concern the furniture and the kitchen. As regards the furniture, it would be useful if we could have an inventory of the furniture in the Lodge at present, showing what is College property and what belongs to Sir William and Lady Hawthorne and will no doubt be wanted by them in their new house. We can then decide what is needed and whether it can be spared from our present home (which of course we will be keeping), what we would like to buy and whether the College could lend us, at least temporarily, whatever else is needed.
>
> Then there is the question of the kitchen, the arrangement and equipment of which certainly looks very dated and not altogether practical for either our normal way of life or for the entertaining we want and expect to do. A good start in this direction would be if we could have a plan of the kitchen and a list of the equipment. I should say that I feel the kitchen is a place where we certainly would like to make major changes.[46]

On 14 June 1983 Hywel George presented a paper to College Council:

> The Master-designate will not be moving his personal furniture to the Lodge until September 1984 and arrangements will have to be made to furnish the main rooms on a temporary basis until then. After discussions between Miss Hammerton and Lady Bondi, details of the furniture required have been agreed on the basis that most of the furniture will be available for use in due course in other parts of the College or will be purchased by the new Master. Total cost of furniture including VAT is £3,250.[47]

The Council approved the proposal. The basic furniture and equipment was duly purchased and a Rational 86 Light Oak kitchen was subsequently fitted, providing Sir Hermann and Lady Bondi with a kitchen which better met their needs and plans for the entertainment of College guests.

During that first year, Sir Hermann and Lady Bondi remained in their home in Reigate. Sir Hermann was still working in Swindon and Christine was teaching in Reigate. However, they travelled to Cambridge most weekends and many of the arrangements were made during these visits. The Domestic and Conference Manager, Anne Hammerton, helped Sir Hermann and Lady Bondi with their move to the Master's Lodge. Anne remembers this very clearly:

Sir Hermann and Connie the cat (1992)

I first met them over the May Bank Holiday in 1983. I was involved because the Housekeeper, Mrs Champion, was on sick leave. The intention was that they would move their own furniture to the Lodge in a year or so's time, but until then any items that had to be purchased would be items that could be used elsewhere in College. We got hold of a lot of catalogues, but a lot of the items we purchased were somewhat ordinary. They were insistent about this. I remember in particular the Ladderax bookshelves that lined the top landing. The items in the kitchen were also basic but serviceable. It was like setting up a new home.

The third Master and his wife finally moved into the Master's Lodge on 19 September 1984, forty-seven years to the day in 1937 when Sir Hermann first crossed the Channel to England. However, despite the College's best efforts, there was one arrangement that was not quite in place. Although the roof of the Lodge had been leaking badly for some time, for various reasons repair work did not start until August 1984. As the work entailed the removal of asphalt and the construction of a new timber framework, the contractors, Rattee & Kett, had already placed heavy tarpaulins over the roof in preparation. Work on stripping the roof was therefore well under way and Christine recalls that this meant that if it had rained over the weekend, when the tarpaulins were removed on the Monday morning, water would cascade into the Lodge. Her recollections of those first few months in the Lodge are therefore somewhat clouded by visions of the buckets that had to be placed all over the top floor of the Lodge to collect the leaks.

Inevitably a brief period of adjustment ensued, not only for the Bondi family but also for the College. The College staff in particular were a little in awe of having as their Master, a man of such distinction. Alice Bondi recalls:

> I was there for the first Xmas in the Lodge and for reasons I don't now recall, two of the catering staff appeared regarding, perhaps, wine for our Xmas dinner. My jaw was somewhat dropping at the deferential manner of these two men – quite unfamiliar to me! – but I really had to work extremely hard not to burst out laughing when they bowed and made to exit the room by walking backwards, as if my father were royalty – and went out of the wrong door, into the dining room rather than the hall. Once they were gone, I am afraid I collapsed in hysterical giggles.

Notwithstanding this a comfortable working relationship soon ensued.

An inventory of the Master's Lodge dated 1 March 1985 provides a delightful snapshot of the furnishing of the Lodge in the 1980s.[48] At that time the Lodge had three bedrooms and a self-contained flat on the first floor (for a resident Housekeeper), as well as a kitchen with a set of servants' bells.[49] In addition to the formal furniture which formed

Sir Hermann and Lady Bondi in the kitchen of the Master's Lodge (1989)

part of the Bracken Bequest, however, there were some more contemporary items. For example, Sir Hermann's study housed a large satinwood desk, four tubular chairs, a gold carpet square and a small standard lamp with orange shade. The Dining Room, which in those days was the room adjacent to the kitchen overlooking the small enclosed garden, had a large Indian carpet with a pale gold background with floral design, while in the hall there was a South African red leather chair with open arms. The 'secretary's office' had a blue carpet square and in the sitting room, there was a Parker Knoll Norton Range three piece suite, a glass-topped coffee table (30" square) and two wooden standard lamps with hessian shades. Years later, I recall that many of the polished wooden floors throughout the Lodge were carpeted in the sort of long pile cream coloured carpet that was so popular in the 1980s. Most evocative of all, though, was the distinctive smell of the Lodge and the Cockcroft Room that adjoined it; a smell of teak, polished floors and cleaning fluids. It is a smell that on occasion still pervades today.

When I moved into my own small office in the Master's Lodge in 1988, I inherited a smart office with a good quality desk and bookcase in the Ercol style and built-in white-painted cupboards which spanned one wall of the room. On the mottled marble slab which covered the heating system and doubled as a window seat – so ubiquitous in College – were two neat wire baskets: one for Sir Hermann's incoming mail and the other housing papers to 'bring forward' for Sir Hermann's daily appointments, while the cupboards housed box after box of carefully indexed reprints and papers that Sir Hermann had written, to be sent to anyone who requested copies (which I seem to recall happened on a regular basis). These were carefully listed as follows:

Box 10 –   Box files – Sir Hermann Bondi
            Reprints Box 3, 1974, plus 2 boxes photographs

Box 11 –   Box files 4-6, 1975-77
            Sir Hermann Bondi Reprints

Box 14 –   Box files No 13 and 14, 1983, 1984
            Sir Hermann Bondi Reprints and Monthly
            Annual Reports[50]

The Master's Office in the Master's Lodge (1990)

I note from a list of the files I sent to the Archives Centre in August 1990 that in addition to these, other files included Ministry of Defence (Unclassified), old correspondence relating to broadcasts, Parliamentary and Scientific Committee (1988-90), Naval Radar, summary of lectures, Severn Barrage Committee, as well as numerous College and University-related papers. With the exception of a file I inherited from my predecessor entitled 'Very Miscellaneous' (from which I deduced she had encountered some difficulty in classifying certain documents), the office was a model of good organisation and efficiency.

By way of complete contrast, Sir Hermann's study could only be described as chaotic. Pride of place on the satinwood panelled wall was the Einstein drawing by Eugene Spirg of which he was so proud. This had been a gift to his father from a grateful patient and it was something he valued greatly.[51] He had an extremely large desk in the centre of the room which was strewn with papers. Once and once only, I tried to tidy a corner of that desk and it was to be one of the very few occasions I can remember being admonished by him. I soon realised that his desk was his filing system and that he knew exactly what was on his desk and where to find it. Two years later, as he and Christine prepared to move out of the Master's Lodge, I remember sitting at my typewriter one day and being startled by a whistling sound emanating from Sir Hermann's study. I peered around the corner and saw Sir Hermann standing there. It was only then that the penny dropped. He was filing. The whistling sound was the sound of a document being skilfully piloted towards a pre-determined corner of the room. And I can still hear Sir Hermann explaining to me in his deep accented voice that it was like geological layering: the papers were being unearthed in chronological order.

Sir Hermann was a delight to work for. I suppose it helped that I was not inexperienced in the role, having previously worked for two former Masters of Emmanuel College, Sir Gordon Sutherland and Professor Derek

Brewer. But much as I had enjoyed my previous jobs, this was somehow different. It was not just the fact that Churchill was a new College in a modern 1960s setting which somehow made everything around it seem outward-looking and innovative; and I don't think I was in awe of him either. I think it was more the fact that despite his great intellect and achievements he was quite simply a thoroughly nice man: never demanding or impatient with me, but always courteous and polite. As Jack Miller once said:

> Achievement in Cambridge is often accompanied by a certain arrogance, but not Hermann, his straightforward approach to problems and his humility when dealing with people resolved many tricky situations.

I would often come in to work on a Monday morning to find that Sir Hermann had spent the weekend writing what was to be yet another paper for the Royal Society or other learned publication. He would come rushing into my office eager to tell me about the task in hand, to dictate a letter or to discuss the day's business. I would then transcribe the paper from Sir Hermann's handwritten notes, always remembering to leave spaces for the algebraic formulae to be added at a later date, having neither the understanding nor the technical capability to do this myself either electronically or manually. Phone calls would then be made; letters would be typed and enclosed in smart white envelopes, immaculate carbon copies being filed away in the appropriate file. The post would then be placed on the tray in the hallway for collection by the Porter on one of his twice daily visits to the Master's Lodge. Life somehow seemed much more courteous and urbane in the days before computers and email.

Many a time I would walk into the study and find Sir Hermann working at his desk with Connie, the cat, who would be draped around his neck like a scarf. Connie was a fully grown short-haired ginger cat. There was nothing timid about her and she lived a happy life interrupted only when VIPs visited the College and sniffer dogs would come into the Master's Lodge to do their duty - which appeared to include the requirement to finish off the food in her cat bowl in the kitchen as if by right. She was greatly troubled and upset by these outrageous intrusions into her otherwise contented life, but seemed at her happiest accompanying Sir Hermann while he worked. According to Christine, Sir Hermann used to say that the traffic hump on the road outside the Master's Lodge was actually the 'Connie bump', possibly because Connie might have otherwise got herself killed on the road had nothing been done about it. Although she was once discovered in a student room on staircase 5 by one of the bedmakers, Connie otherwise managed to live her life in a manner appropriate to a Master's cat. She moved to Impington with Sir Hermann and Lady Bondi in 1990 where she lived out her final few years in relative contentment.

Despite the inevitable teething problems, however, Sir Hermann and Lady Bondi soon began to settle into life back in Cambridge and the Master's Lodge was to become very much a home to them. For some years, Brenda Pledger, the Master's Lodge Housekeeper and wife of the Deputy Head of Grounds & Gardens, Graham Pledger,[52] came in every day to clean the Lodge and for some of this time also, Sir Hermann and Lady Bondi's youngest daughter, Debbie and her then husband Andy, lived in the self-contained flat with their new baby, Ben, (although ever thoughtful of others, when the flat was vacant, it was lent out to Visiting Fellows, once even being used by the wife of an Overseas Fellow as a small studio in which to do her acrylic painting). When the grandchildren visited the sounds of laughter and giggling would inevitably fill the Lodge. Debbie recalls:

> From around October to December 1988 (whilst Andy and I were living in the flat), I went back to work in London, leaving early in the morning to catch the train. As a result, Christine looked after Ben from the time Andy went to work to the time that Rachel Butler [the child minder] arrived and then looked after Ben at the end of the day when Rachel went home. So for a while, the kitchen in the Lodge became a focus for Ben, Rachel, Christine and Brenda who always found time to come to see Ben.

Sir Hermann with grandsons Tom (in picture) and Max (1985)

Christine also recalls that two of her grandsons, Max and Tom, then aged about six and eight years, used to love visiting the Lodge as there were so many hiding places and play areas. One of their favourite places was the space under the main stairs where they would have tea parties. They also discovered that they could get under the structure of the house and they spent much time playing under there. Debbie and her family moved out of the Lodge some time in 1989. Her daughter, Claire, was born in 1990 and I remember Claire's first visit to the Lodge and sharing with Ben his excitement at his new baby sister.

When the grandchildren were not there, however, the Lodge was often silent, save perhaps for the sound of Sir Hermann speaking on the telephone, or Christine and Brenda Pledger chatting quietly in the kitchen, or the sound of the doorbell ringing briefly (to announce the arrival of the Porter collecting or delivering the mail). Even today I can still hear it: the clunking of the door being unlocked, a couple of footsteps, a flapping sound as the post was dropped on the table and then the door being locked again and all within the space of about 20 seconds before the Lodge was engulfed in silence again. But I grew used to the silence of the Lodge and found that this way of working suited me well, though it came as quite a shock when Alec Broers became Master in 1990 and installed in the study a rather sophisticated hi fi system and Beethoven or Mozart would echo throughout the ground floor of the Lodge. This was a change which I readily embraced and loved, save for the difficulty of answering phone calls professionally under such difficult circumstances.

Sir Hermann's various passions inevitably pervaded his everyday life. If his appointments schedule would permit, he would leave his study mid-morning, go to the kitchen and shut the door. I would then know that he and Christine were having morning coffee together. Sometimes they would be chatting but at other times I discovered that they would be doing a giant jigsaw puzzle together on the kitchen table. Sometimes, if the jigsaw was particularly complex, they would use a table upstairs and work on it over a period of days. On more than

Sir Hermann and Lady Bondi in front of a portrait of Sir John Cockcroft, the first Master (1984)

one occasion I remember Sir Hermann ushering me upstairs to see the completed puzzle, such was his pride at having completed so complicated a task.

Sir Hermann's passions are clear in the recollections of others. For example, a former By-Fellow, Wolfgang Rindler, recalls a visit to the Lodge:

> ... he had an incredible collection of maps and guide books, essentially all the Ordnance Maps covering the British Isles, but also all of Europe was covered and, for all I know, the other continents too. Mention the smallest alpine village in Switzerland and a minute later Hermann would fish out just the right detailed map of the area. Other people would at least rummage for a while in old boxes before finding that. Not Hermann! I don't know what his system was, but the result was almost instantaneous. I suppose the superior system was just his superior mind. He had a passion for alpine railways and the daring engineering that made their construction feasible. He especially liked and knew all the tunnels. Especially those that doubled up on themselves within the mountain! You felt that if pressed he might even come up with the time the trains ran. The British Ordnance maps he used for hiking and archaeology.

Reading this, Alice Bondi comments that her father would have come up with the timetable for the major train routes through Europe – he apparently had them all in a big annual guide.

Debbie Bondi recalls the 'port railway' in the dining room of the lodge:

> ... which Hermann proudly showed off to the amusement of my siblings. It was kept in one of the cupboards in the dining room of the Lodge. There was a wooden track which fitted together into a circle or an oval – it was clear that it had been made to go on a dining table rather larger than the

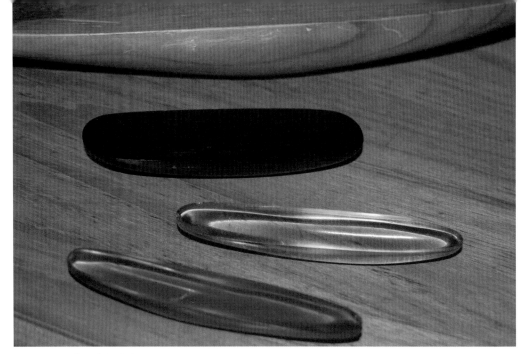

A selection of 'rattlebacks'

one in the Lodge. I think there was a single 'carriage' which ran on the rails which carried the bottle of port around the table. I never saw it used for its intended purpose – I just remember Hermann's childish excitement at anything to do with railways!

The Domestic & Conference Manager, Anne Hammerton, remembers some of his other 'toys' which included a piece of wood which he would spin. Christine Bondi says that this was a 'rattleback': it was not totally symmetrical and could be pushed in certain ways. She says that Sir Hermann would love to produce these for visitors, particularly when there were students around.

During these years Christine found that her own workload continued to be demanding. Notwithstanding the social demands of being the wife of the Master, Christine continued to pursue her own academic and other interests. Rather than concluding her own teaching commitments in Reigate in 1983, she chose to continue through to 1984 and, like Sir Hermann, was only able to visit the College at weekends:

After we had moved back to Cambridge, I first of all did some teaching at the Perse School for Girls for three months or so from December 1984. I was also editing a maths book for the IMA (Institute of Mathematics and its Applications) and this took up a lot of my time. In the spring of 1985 I was standing for Histon and Impington (before we lived here) and the next year I was asked to stand for the City Council. I nearly got in, but lost by 4 votes. Thank goodness I didn't get in, I would have hated it! I was persuaded to stand again in 1987, following John Brackenbury's departure, but I knew I wouldn't get in as you really did need a local candidate.

A major aspect for me was also our visits to India, the first being in December/January 1985/6. Hermann gave lectures all over India and I managed to see lots of schools and places like the Nehru centre in Mumbai. It was all under the auspices of the British Council. We went back to India, mostly at the invitation of Humanists and Atheists in 1987. As a result I saw the work supported by *Save the*

Children in Andhra Pradesh and as soon as I got back I got in touch with SCF here in Cambridge and got involved rapidly. We had some events at the Master's Lodge, including coffee mornings and one or two cream teas on the lawn. I also became involved in reviving the Cambridge Humanist Group about the same time.

Later on, Frank King, who was organising a University course in Elementary Maths for Biologists, invited me to come in and do the calculus bit. This meant I had to give twelve lectures in the Easter Term. The first one was in 1989 and I did them for two further years, even after we moved to Impington. It was quite an experience as I suddenly discovered that there were science students in Cambridge who didn't know anything about maths.

Those of us who worked in the Lodge fully respected Sir Hermann and Lady Bondi's need for occasional privacy in such a public life. In March 1985 Sir Hermann felt moved to issue instructions regarding telephone calls:

If a caller asks for the Master, please ask whether he would like the Master's Secretary. If yes, put through to 226 **on weekday afternoons 2.30 – 5.30 p.m. only**, otherwise request the caller to call again between those times. If the caller does not want the secretary, but the Master personally, put the call through to 228.

Without such modest safeguards, the Master and his wife could find themselves expected to be on duty twenty-four hours a day.

Sir Hermann and Lady Bondi talking to Albert Richmond, the Head Gardener (1984)

Sir Hermann and Lady Bondi in the Master's Lodge garden (1984)

PRACTICAL SOLUTIONS

There is occasionally a little trouble in the bathroom when one is showering.[53]

Sir Hermann took an active interest in seeking practical solutions to problems. I would often come in to work on a Monday morning to find him ready to dictate a memo or letter to report a fault, a failure of a system somewhere, or mid-way through writing a report that was destined to offer a proactive solution to a problem. This was certainly not due to any irascibility on Sir Hermann's part: it was simply that he felt strongly that matters should be put right. If he could offer a solution, he clearly felt that it was his duty to do so. He articulated his particular approach to problem-solving in an article he wrote while Chairman of NERC, in which he said that he had always thought that as his background had been intellectual, anything practical would be beyond him. He had compared this view to the attitudes of Sir Fred Hoyle, 'with his down-to-earth outlook' and Professor Tommy Gold 'with his engineering background and interest':

> As he (Tommy Gold) put it bluntly, "If you have the analytical intelligence to examine an astrophysical problem in fluid dynamics, turn it into partial differential equations and solve them; then that same analytical intelligence can be exercised just as severely and successfully to analyse and correct a fault in a hot water system or in an old car." This was a revelation to me, particularly as it worked! (Ten years later, when I ran an old car, I soon discovered that a good garage mechanic, with his experience, could spot a common fault much more quickly than I with my ponderous analytical approach, but, if it was an unusual fault, I beat him by a handsome margin.[54]

As far as problem-solving within the College is concerned, the College is most fortunate in having expertise available within the Fellowship on which it can call for advice on how to deal with problems that may require specialist input. Inevitably the down-side of this is that the Bursar can occasionally find that he or she faces a surplus of sometimes conflicting expert advice which can make the solution even harder to determine. Fortunately, at other times problems are easier to resolve.

Anyone who has ever experienced communal living will know that there are both advantages and disadvantages to such an arrangement and for the third Master and his wife there is little doubt that living over the shop had its fair share. For example, on 18 May 1987 Sir Hermann wrote to the Head Porter, Peter Bullock, as follows:

> We are having some troubles with the front gates to the Lodge. The gates are often locked, though not consistently, when we are away. One difficulty is that my wife's key does not operate this lock (though mine does). When they are found locked late at night in pouring rain (as last night) this is very aggravating. This problem could be solved by changing this lock so that my wife's key operates it.
>
> However, this does not deal with the other difficulty in that certain deliveries (evening paper and mail

The Churchill swimming pool (c. 1982)

which is privately delivered) are generally made to the front door. To find a sodden newspaper stuck through the locked gate is not only irritating but no contribution to security.

I very much understand that you must have security very much in mind. Therefore I would appreciate your advice.[55]

Security and access issues apart, it is hardly surprising that with so many people living on site and conference business growing as rapidly as it was in the 1980s, there would also be problems with unauthorised parking on the Master's Lodge forecourt. The car park at the top end of the College was not destined to be built until the 1990s and according to Peter Bullock there were just 80 parking spaces for visitors and students. It was therefore inevitable that at busy times the empty spaces on the Master's Lodge forecourt would become a magnet for unauthorised parking. For example, on 3 March 1986 Sir Hermann wrote to the Domestic & Conference Manager, Anne Hammerton:

We suffer, at times most irritatingly, from unauthorized parking on our forecourt from people attending conferences. At the same time we would wish to help rather than hinder our conference trade which is so important to us and of course we rarely need the whole of the forecourt. I would therefore make the following suggestions:

i) If a conference is likely to involve a significant number of cars to need parking but there is no shortage of parking space further up the Private Road, 'No Parking' cones are put along our forecourt edge.

ii) If there is likely to be a shortage then, after consultation with us, the part of our forecourt West of our entrance is given a label 'Conference Parking' while the part to the East is given 'No parking' cones.

iii) If we are away and assured space is left for our domestic help and for my secretary to park their cars, all the rest of the forecourt may be used for conference parking.

I appreciate that this will involve extra work for the porters, but it should help our Conference Parking and avoid a serious nuisance to us.[56]

Anne Hammerton's response to this memo is not on file, but she says that Sir Hermann was always incredibly supportive of conferences. He would invariably receive many requests to address conferences and would always try to do so if his diary permitted, requesting a copy of the conference programme and other details so that he was fully informed before the event. Fortunately this meant that a small matter such as shared parking on the Master's Lodge forecourt could be easily resolved.

But there were other drawbacks to living in College, not the least being problems with the central heating and plumbing systems in the Master's Lodge. These were to prove more difficult to resolve. Indeed, Christine says that they never did manage to gain control of the central heating and although the Chief Maintenance Engineer, Sid Brown, did show her how to operate the system, she says that she never felt confident enough to run it properly. As a consequence, Christine says that she felt that they were wasting a lot of heat at times, which must have been particularly frustrating for the former Chief Scientific Adviser to the Department of Energy.

In August 1985 Sir Hermann, wrote to Sid Brown, the Chief Maintenance Engineer:

> If you would like a restful quarter of an hour one day, I suggest you operate some of the equipment in the bathroom and then come downstairs to the lounge and listen for a quarter of a hour or thereabouts to the clicking noises, the dripping noises and the like. We have learned to live with this very minor embarrassment by not using our own bathroom (there are plenty of others) when we have guests in the lounge. On the other hand, it does suggest that plumbing in that wall is extremely peculiar and may at some stage give trouble.

> Secondly, there is occasionally a little trouble in the bathroom when one is showering, with the water running down the shower hose or otherwise, escaping past the curtain, to make a puddle on the floor between the bath and the toilet. If it were easy to fix a piece of, say, Perspex 6" x 8" wide from the rim of the bath to the ceiling, then this would keep the shower curtain well in this area and avoid this kind of trouble.[57]

On 21 May 1985 the Buildings Committee considered a paper by the Master entitled 'Heating System and Domestic Hot Water Supplies'. Minute 680 confirmed the action agreed:

> The Committee took note of a paper by the Master and approved the proposal that full drawings should be made of the location of the pipe runs and valve positions in the Master's Lodge area and North Court (diagrams of the system in the remainder of the College were already available). The Master was authorised to arrange this and it was agreed that Mr Tizard should be invited to set out the complete control system in a proper engineering manner.[58]

Notwithstanding the College's fortunate position in having access to expertise within its Fellowship, these particular problems were destined to continue throughout the third Mastership. In January 1990 Sir Hermann was moved to write to the Bursar, Michael Allen, following a boiler breakdown on New Year's Eve. Typically, his concern was not so much at the inconvenience to himself, as to possible inconvenience to conference visitors. He wrote:

> When we are not here, such a breakdown can go unreported for a long time causing great inconvenience to all who live in College, including conference visitors. Should we not have an automatic monitor ringing in the Porters' Lodge whenever, say, the water temperature drops below a pre-set level?[59]

The final letter in the file is again to Michael Allen. Sir Hermann wrote on 9 April 1990:

> I would like to have a discussion on the future of our system. While I entirely appreciate that, with heavy oil as a fuel, centralization was right; I find it difficult to believe that if gas is to be our chief fuel, decentralization is not preferable.[60]

The Master's Lodge garden in summer (1984)

Clearly the former Chief Scientific Adviser now meant business.

From the beginning of his Mastership, however, there was another watery matter weighing heavily on Sir Hermann's mind and this was the desirability or otherwise of the swimming pool which was located just outside his lounge window.

In June 1978 the College Council had approved a request by the Master, Sir William Hawthorne, for permission to build a swimming pool in the garden of the Master's Lodge, the cost being borne entirely by Sir William. Mary Beveridge says that Sir William used the pool regularly. She recalls that members of the Boat Club used to go swimming in it after bumps suppers and that the pool was well used.

Anyone with experience of maintaining a swimming pool, however, will be aware that there is a price to pay for the simple pleasure of access to one's own pool and that is the not insignificant effort involved in its maintenance. Some find this effort tiresome; others find it a challenge they are willing to meet. If one is busy or travelling overseas, however, this will occasionally require the assistance of others. Fortunately, a College is an ideal place in which to site a pool as it has a vast supply of qualified and willing helpers. For example, on 7 November 1980, Sir William Hawthorne wrote to a student:

> I am in urgent need of your assistance as a frogman in my swimming pool. In fact I have put a little heat into it so you won't feel too cold.[61]

While in 1981 he was apparently looking for a 'swimming pool sitter' as an undergraduate student, M H Clements, offered to take on this task.

But there is no doubt that the daily care and maintenance of a pool would be a burden to one who had no desire to make use of the facility. Mary Beveridge says that Sir Hermann was particularly worried for the safety of his young grandchildren and very early on in his Mastership he made clear that he would prefer the swimming pool to be filled in. In 1984, in an undated letter to Mary Beveridge, he wrote:

> With the better season coming, decisions are more urgent. We would be very happy for the pool to be used by senior members of the College and their families provided none of the work of keeping the pool in order falls on us.[62]

But the College already had its own concerns. Hywel George wrote to Sir Hermann on 27 April 1984:

> I am a little nervous about allowing the swimming pool to be used when the Lodge is unoccupied. When you are away the front gates are locked and allowing access will not help security. There is also the problem of safety if children are allowed to use the pool. Perhaps we can discuss this on your next visit.[63]

The matter was discussed by the Buildings Committee on 21 May 1985. Minute 679 stated:

> The Committee was informed that the Master had said that his interest in the swimming pool was minimal and that it should be "mothballed". The Committee expressed concern, since the pool had been a gift from Sir William Hawthorne to the College, but it was noted that the Master had discussed the matter with Sir William. It was agreed that and if the Council approved the Master's request, the work should be done by covering the walls and floor of the pool with polythene sheeting and filling it with sand. The surface could be covered by soil, turf or paving slabs and the Committee agreed that soil should be used so that the area could be made into a flower bed.[64]

The College Council approved the proposal at its meeting the following week. Maintenance costs and Health and Safety implications alone would have made this a prudent decision in the circumstances. The swimming pool was duly mothballed and remains so to this day.

Sir Hermann with grandson Ben at home in
Impington (1993)

FEASTS AND FORK LUNCHES

It seems to me really not a great demand that there should be at every place a water glass and plenty of jugs of water distributed over the tables. It is a normal civilised habit around the world and I do think we could follow it.[65]

An important duty of the Master of any College is that of entertainment. This may include the entertainment of distinguished visitors and senior academics and attending High Table, through to hosting receptions for students and staff and the Master is provided with a modest allowance to enable him to perform this task in an appropriate way. Although the Master's wife does not hold a formal position within the College, she will inevitably be required to play a full role in social engagements and other events, whether by accompanying the Master to High Table or other events in Hall, attending High Table in her own right, attending receptions in the Master's Lodge or Cockcroft Room, arranging and hosting dinner parties, as well as looking after College guests staying in the Lodge. Although this would regularly encroach on their weekends, this was a task that Sir Hermann and Lady Bondi embraced fully, and arguably the one that gave them the most pleasure.

Christine recalls that they used to invite the students for a 'fork lunch' in the Cockcroft Room and that they would try to gather together an interesting mix of guests to ensure a flow of good conversation. This would often include the neighbours on Storey's Way, as well as members of staff, the inclusion of the latter, according to the Domestic & Conference Manager, Anne Hammerton, having been unheard of until that point. The first-year students also used to be invited to a light afternoon tea where sandwiches and cakes would be served, while in later years students would be invited for summer drinks in the Master's Lodge lounge or to a 'strawberry tea'. If the weather was fine, the doors would be opened and the students would spill out on to the patio. Christine says that the family gave them a gift of an extremely large tea pot to use on such occasions and this was often used at these events, the pot itself being donated to the MCR when they moved out of the Lodge.

The former Vice-Master, Jack Miller, wrote:

> Hermann was half of the team. He and Christine spent much time entertaining our junior members in the Lodge where they saw for themselves that in Churchill barriers just do not exist. We all have much to thank them for.

Jack Miller also recorded that he had been particularly impressed by the way in which Sir Hermann put post-graduates at their ease by treating them as equals. In addition to this, Sir Hermann and Lady Bondi were particularly good at circulating at such events and would ensure that they moved around the room frequently, chatting to everyone they saw, Sir Hermann inevitably regaling guests with stories or giving advice in his own inimitable way.

Wolfgang Rindler, former By-Fellow, recalls that Sir Hermann had a huge reservoir of jokes and stories, 'easily matching almost any occasion when needed'. He says:

Sir Hermann at Sid Brown's retirement party (1989). Sid was Chief Maintenance Engineer from 1962-1989

Of course, every one knows what a great conversationalist he was, how carefully and precisely he chose his words, how he wielded the English language like a surgeon wields his scalpel. He must have completely interchanged his original German (we were both born in Vienna) for English when he came to England and his beloved Cambridge. I don't think he ever cared to speak German again.

A former Fellow, Mark Tester, recalls some advice he was given at a social event when his own daughter, Gabriella, was just starting to walk. Terming this 'mobility without responsibility', Sir Hermann had commented that it was a particularly interesting stage of life. He had also recounted that when he had a young family, it had been much easier for him to get into the playpen and write his papers, leaving them to roam the house, rather than have them in the pen wanting to get out. And at other times, Mark recalls that Sir Hermann would be seen looking into his glass and muttering something about how hot his hands must be that evening since the wine kept disappearing from his glass. Mark says that Sir Hermann's own conclusion to this particular phenomenon was that the wine had to be evaporating.

A duty that Sir Hermann and Lady Bondi particularly enjoyed was that of attending High Table as regularly as their respective timetables would permit. High Table is the table reserved for Senior Members of the College and distinguished guests. It is the long table on the left hand side of the Dining Hall as you approach from the main staircase. It provides an opportunity for the Fellowship to network, to build new academic liaisons, to entertain VIPs, to promote the College to potential donors, or simply to get to know each other. The former College Secretary and Registrar, Mary Beveridge, recalls:

Hermann being a humanist was reluctant to say grace before High Table or any other College dinner. He was persuaded (I think by Jack Miller) that saying "benedictus benedicat" before and "benedicto benedicatur" after had little religious significance and was unlikely to compromise his humanist principles. But I don't think he was ever persuaded to use the long Latin version of the Churchill grace.

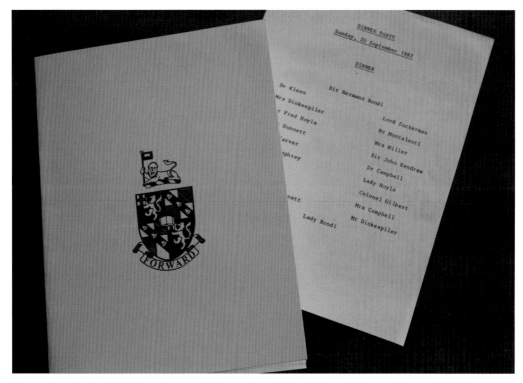

The seating plan for the Dinner Party on 20 September 1987 which was held to celebrate the 50th anniversary of Sir Hermann's arrival in the country

Notwithstanding this slight difficulty, which Sir Hermann managed to surmount in his own inimitable way, he and Christine took particular delight at these events in talking to the senior academics who were invited to the College under the Visiting Fellowship programme. They were also generous with their time, their interest often moving beyond that of polite conversation. Wolfgang Rindler writes:

> Hermann and Christine took us for some wonderful walks in the countryside and gave us great hints on abbeys and other antiquities to visit … Along with Christine, Hermann was the most delightful and considerate of hosts.

At weekends they would also sometimes invite the Visiting Fellows to join them on trips to Anglesey Abbey or other historic houses. Banmali Tandan, a former By-Fellow, remembers these trips very clearly, particularly since he would occasionally be invited to join them on his own. He says that the three of them also spent many days together on travels within India during which they had many memorable experiences. While Jonathan Parker, a former advanced student, recalls that in addition to visits to the Master's Lodge, Sir Hermann once invited him to tea in his home just outside Cambridge in order to meet his children. Jonathan admits that as MCR President he had the privilege to get to know him better than maybe otherwise but was nonetheless grateful for the helpful discussions he was able to have with Sir Hermann about his own research on environment and energy policies and very much enjoyed these occasions.

Wolfgang Rindler also recalls Hermann's academic generosity:

Sir Hermann christening a new Churchill boat (1986)

When I later spent an Easter Term at Hermann's invitation as a By-Fellow at Churchill, in 1990, I immediately got a taste of Hermann's generosity. Already by mail he had suggested a very interesting problem for the two of us to work on. But for one reason or another, my trip got delayed and by the time I got there, Hermann had essentially done the problem and written the paper. All that was left for me to do was to check the calculations, for which a tiny note of thanks at the end of the paper would have been ample reward. But Hermann would have none of that and insisted on publishing the paper in both our names!

Back in College, however, the Domestic and Conference Manager, Anne Hammerton, was probably the first member of staff to experience at first hand Sir Hermann's propensity for extremely weak tea:

For their first (temporary) stay in 1983 we put provisions into the Lodge for them. Sir Hemann insisted on making the first cup of tea in the Lodge and would not let Lady Bondi do it, although she was able to warn me about the weakness.

Jack Miller commented on this aspect also:

Hermann's tea ceremony after dinner was a class act. Disliking coffee, he would be served with a pot of hot water and a tea bag. The bag would be lowered into the water for a count of five then squeezed by hand. He had the dining room staff well trained in this procedure.

Very early on into his Mastership, the catering staff grew adept at serving tea at the correct strength. This did not go unnoticed by Sir Hermann and in a memo to the President of the SCR and the Fellows' Steward, dated 28 May 1987, he wrote:

FLYING ROAST DUCKS

I am most appreciative of the courtesy that has been shown to me since I arrived in weak tea being available to me whenever I am dining or otherwise taking part in a function. Nowadays it occurs so frequently that others would like to have tea after dinner that I wonder whether it would not be possible to extend the choice generally, by having a pot of tea and a jug of hot water available normally.

So weak was the tea, however, that it was to become a focus for discussion. Chris Smick was a Schoolteacher By-Fellow in 1993 and continues to have a close association with the College, returning every summer. His electronic diary reveals the following:

Wednesday 25 May 1993
It was either during the SCR conversation, or on the way up the Fellows' stairs that Keith [Williams] told me the name for Sir Hermann's malady, aka his tea deficiency. It is called "Ateaosis" (pronounced "eh tea osis").

Friday 12 August 1994
While I was talking to John [Atiyah, at that time the College's Computer Manager], in his office, Sir Hermann Bondi came in to see him. I politely rose to greet him and reminded him of who I am. When I mentioned the mileage I've gotten from the disease we share, he told me that Richard Adrian had told him that the name should more properly be "hypoteamia".

Thursday 27 April 1995
Dined with [Christine Northeast]. (I was at the end of the table, Sir Hermann was presiding). After dinner conversation about ateaosis with Sir Hermann – I asked him if he knew the name for the analogous acute shortage of alcohol, but he didn't..

These recollections would have greatly amused Sir Hermann who would have been pleased to hear that in the institution in Massachusetts where Chris still works, 'hypoteamia' is now used in its adjectival form by staff in the staff room when they are feeling somewhat 'hypoteamic'.

Sir Hermann was to preside over many College Feasts and other formal College dinners. He always enjoyed these and his speeches were appropriate and well received. Fellow and Vice-Master at the time of writing, Alison Finch, writes:

At dinners for the students, particularly those celebrating the Founder or in some other way significant for the image of the College, he would make excellent speeches reminding all present of the moral purpose that had contributed to Winston Churchill's greatness as a war leader and of the dreadful evil against which Churchill was fighting – something current students maybe aren't always fully aware of. Hermann's public face not only as a distinguished scientist but also as a leading representative of British secularism also seemed to many of us to exemplify the spirit in which Churchill College had been founded. One of my last memories of him is of a short but characteristically brilliant talk he gave at a conference to celebrate our French Government Fellowships. As ever, he was pithy and memorable.

There were also some benefits for the former Vice-Master, Jack Miller, who wrote:

Once Hermann's appointment here had been announced, I spoke to one of my contacts in NERC to seek 'further and better information'. I was told that Hermann enjoyed travelling and speaking. The latter was to prove very useful. When Will Hawthorne was Master, he and I would take it in turns to chair a big annual conference dinner.[66] We had devised a speech which was more or less the same each year. The dinner was held on the first Sunday of the long vacation when sitting with one's feet up would have been preferable. Once Hermann heard that a speech had to be made, he was in there like a shot. Result, we were both happy.

Jack's widow, Marcia, also recalls that Sir Hermann was the first Master to get to his feet and speak after a College dinner. She can still recall Jack's sharp intake of breath when Sir Hermann did so. There were other exciting occasions, too, not the least being when Sir Hermann accidentally spilt a jug of water over Denis Healey. And those who sat next to him at dinners and feasts often had fascinating discussions. Bill Barnett, a current Fellow and the former Keeper of the Archives Centre, recalls such a conversation:

At a High Table dinner after Hermann had retired as Master, my guest and I were discussing why in the 1970s Jim Callaghan's Labour Cabinet had opted to develop the British 'Nimrod' Mk III airborne early-warning radar system (eventually aborted as technologically unworkable) instead of the already proven American AWACS system.

I knew that Hermann had been Chief Scientific Adviser to the Ministry of Defence at that time and there he was sitting next to me at High Table. So I put the question to him and this was his answer:

'I was NOT asked as a scientist which I thought would be the better technical and operational choice. Instead, the question put to me by ministers was: "Can you say as a scientist that the Nimrod system will not work?" Well, with Nimrod Mk III then still on the drawing-board, of course I could not say for certain that it would not work. But that negative answer was enough for the Government to opt for Nimrod. What really mattered to the politicians was not the rival merits of the British and American early-warning systems, but British jobs in the avionics industries.'

Bill suggests that this anecdote illustrates the contempt which Hermann, with his first class scientific intellect, felt for the time-serving politicians.

Sir Hermann and Lady Bondi's youngest daughter, Debbie, says:

I recall a dinner Christine and Hermann had early on in Hermann's period as Master at which Lord and Lady Carrington were guests. I remember it well, firstly because I was still a student in Manchester at the time and I had nothing suitable to wear. I remember somewhat challenging negotiations with Christine about what she would be prepared to pay on a suitable outfit! More noteworthy was the comment Lord Carrington made to me and my brother David at the dinner itself: he reflected on Hermann's incredible achievements having come to the UK as a foreign student, having been interned as an enemy alien etc and that he had achieved so much more than many who had had a far easier time of things.

Christine still has the Visitors' Book which she would ask visitors to the Master's Lodge to sign. Inevitably the signatures included some very distinguished guests, including Lady Colville, Tommy Gold, Denis Healey, Fred Hoyle and the Lady Soames, to name but a few, but there were also those of the many other colleagues and friends who visited the Lodge. The Bondi children and grandchildren would also write their names in the book every time they visited, the signatures becoming easier to decipher as the years progressed. There was even one signature purporting to belong to Connie the cat, paw-printed on 31 December 1989.

A formal occasion: Sir Hermann and Lady Bondi with The Lady Soames, DBE, and Founding Fellow, Dr Richard Hey (1985)

The Lady Soames was of course a regular and very welcome visitor to College, as she continues to be to this day. She would often stay with Sir Hermann and Lady Bondi overnight and a very warm friendship was to develop. A rather poignant handwritten card from her refers to the Founder's Dinner in 1986:

> I am ashamed not to have written sooner to thank you both for having me to stay for that extra special Founder's Dinner. It was such a memorable occasion – and so enjoyable (as always). I loved Hermann's words about my Papa – and it moves me to know his memory and spirit are cherished in the marvellous institution that bears his name.[67]

I remember another occasion, in 1990, very clearly. Lady Soames had attended an event in College, but Sir Hermann and Lady Bondi had been away and had been unable to entertain her. However, they insisted that she should stay in the Master's Lodge and asked me to come in to work early the following morning to organise breakfast for her. I remember arriving at the appointed time to find Lady Soames already in the kitchen happily making her breakfast. She then insisted that I should join her. I recall our conversation that morning as clearly as if it was yesterday. The warmth she exuded was indicative of how relaxed and welcome she felt in the Lodge, despite the absence of her two hosts. It is a moment in time that I will certainly treasure.

Sir Hermann and Lady Bondi also gave many dinner parties in the Master's Lodge. One such occasion was on Sunday 20 September 1987, which was attended by twenty guests. The guests included Sir Fred and Lady Hoyle, Lady Humphrey, Colonel Gilbert, Lord and Lady Carver and Lord (Solly) Zuckerman, as well as Jack and Marcia Miller and Colin and Margaret Campbell. The wines were no doubt recommended by the Wine Steward, Brian Westwood and included a Muscat 1984 and a rather nice Ch. Mouton Baron-Philippe 1970, with a Niersteiner Auflangen

Auslese 1983 accompanying the dessert. The starter was an Artichoke and Shrimp Salad, followed by Roast Rack of English Lamb, Summer Pudding, coffee and dessert, recommended and prepared, no doubt, under the supervision of either Head Chef, Martin Hayden, or his Deputy, Richard Mee. The Buttery & Cellar Manager around this time would have been Malcolm Stephenson and the serving of food almost certainly presided over by one of other of the Dining Hall Managers, Joseph Carberry or Evelyn Walker.

There are some lovely letters on file following the event.[68] These must have been treasured by Sir Hermann as they were not despatched to the Archives on his retirement as Master in 1990, but transferred to his College room. The letters show that the party was a real success. Margaret Campbell wrote on 21 September:

> Colin and I felt most privileged to be with you and your other guests last evening and would like to thank you most warmly for a truly delightful party and celebration. It was most interesting to learn how your other friends had been involved with you both in so many diverse projects. We never cease to be amazed at all you have achieved and the great enthusiasm and interest you both take in everything you do. Churchill is indeed fortunate to have you both.

Another letter was from Lady Humphrey. She wrote:

> Here I am back home safe and sound and still enjoying and relishing every moment and every morsel of that lovely dinner last night. Thank you again so, so much for asking me to join this very special party and for being most comfortably accommodated for the night, too.

> To think that I was sitting with such a distinguished group of people – such a lot of interesting and amusing conversation – and all of us with one thing in common – the privilege and joy of your wonderfully warm and generous friendship.

> How perfectly Lord Zuckerman put into words just what we were all feeling and what was in all our hearts and minds.

While Colonel (Stuart) Gilbert wrote:

> What a splendid party! And how clearly it demonstrates a most interesting life. Radar, astronomy, space, the Barrier, Defense at the highest level. The Severn was missing (although I had a bit part in it in 1927) and then Churchill College. I am sure that it makes a beautiful home for you … A night to be remembered.

Finally, Lady Carver wrote:

> It was a very special party for a very special occasion and I do hope you both enjoyed it as much as your guests did. Thank you so much for including us for the heart warming welcome and delicious dinner. It was good to see you so happily established at Cambridge and it was interesting to meet your other guests as well as to see some old friends.

Christine remembers that the party was a celebration of the 50th anniversary of Sir Hermann's arrival in the country. She says that the guests were colleagues and friends he had come to know through the various positions he had held in the fifty years since his arrival and included those from the Ministry of Defence and the European Space Research Organisation (ESRO). It was indeed a very special event.

View towards the squash courts and the
Henry Moore sculpture (1984)

As Master, Sir Hermann took a keen interest in all matters relating to students. Jack Miller said that he never neglected his role as Patron of the Boat Club: 'Many is the time I saw him on the towpath during the Lent and May races, along with fellow Masters while the others watched their boats being defeated.' And he supported students in other ways, too. On 24 January 1989, as his year as MCR President came to a close, Jonathan Parker wrote to the Master:

> I am very annoyed and embarrassed by the mindless behaviour I see in some students. This includes obscene drunken bar behaviour, blatant sexism and vandalism. There is no excuse for this sort of thing. I have tried to encourage the artistic side of College life, for example by organising the proposed tour of College works of art. But such efforts seem in vain, when last year the Bernard Meadows statue was ripped off its pedestal. The Bursar now tells me it is not going back, for fear of the same thing occurring again. We should not let vandals determine the artistic destiny of our College. We need paintings and sculptures around the College to brighten it up and complement the architecture.[69]

Sir Hermann responded on 13 February 1989:

> …I agree with you that the mindless behaviour of some of our younger students is a serious blot on our existence as a civilized community. I think it might well be worth investigating this more deeply. Undoubtedly young males in groups inevitably present problems but I would not wish to jump to conclusions without a little more knowledge. Questions that could be looked at, by comparison with other years here and with other Colleges, are:
>
> (i)   Are cohorts with special behaviour problems (usually directly due to a very few individuals, but perhaps actively tolerated by many) also distinguished by poor exam performances? Which is cause and which is effect?
> (ii) Is the ratio of men to women as important a criterion as you seem to think?
> (iii) Could we do better by more intermingling of postgraduates and undergraduates?
> (iv) How do we compare with other Colleges with different social composition?
> (v) Would higher prices in the bar diminish the problems, or would they simply drive people out of College?
>
> Finally, how do we get all this studied? It could be part of a Social Science M.Phil thesis, provided other Colleges would not clam up
>
> … This is necessarily an incomplete response to your letter, yet we might think whether both letters, perhaps slightly modified, should have a wider circulation (JCR, Senior Tutor, TAS, as well as MCR).

An agreement was clearly made that the letters would be forwarded to the College Officers. Unfortunately the next letter in the file is from me. The task of conveying disappointing news had been delegated to the Master's Secretary. On 17 March 1989 I wrote to Jonathan Parker:

> I am afraid that although your letter to the Master was circulated to College Officers, it was not felt suitable for more general circulation.

Although things did not progress as far as Jonathan had clearly hoped, his views had been considered carefully by Sir Hermann who had ensured that Jonathan had received a considered response and that the exchange of correspondence had been circulated more widely.

Sir Hermann relaxing in the Master's Lodge garden (1985)

Sir Hermann had a keen interest in student welfare. On 28 May 1987 he sent the following memo to the Fellows' Steward and Catering Manager:

Water: During a discussion in Council last autumn it was, I believe, agreed that whenever junior members are at a major dinner, water and glasses would be available on the tables to avoid thirst leading to overconsumption of wine. It does seem to me that this is a good rule but it was not carried out at this Term's Advanced Students' Dinner. [70]

On 8 July 1987 he wrote to the President of the Senior Combination Room (SCR), Dr Michael Hoskin:

… I also want to mention a matter that rather irks me. This is the lack of provision of water at Feasts, particularly when Junior Members are present. It seems to me really not a great demand that there should be at every place a water glass and plenty of jugs of water distributed over the tables. It is a normal civilised habit around the world and I do think we could follow it. Senior Members may be willing to ask, which of course puts a strain on the staff during the time when they are busy serving. Junior Members when they are thirsty have little choice but to drink wine. If you recall the matter was discussed at Council last autumn but nothing has happened since and I think it is high time that something were done. [71]

In the same letter Sir Hermann also took the opportunity to express his views on the consumption of alcohol by Junior Members. He continued:

I am particularly concerned over the Freshers' Dinner. So many of our new young people have had little or no experience of wine. It is nice to show that the College is generous and that their glasses are refilled but I think it would be better still, first, if there were water available and, secondly, if the refilling were not quite as excessively quick and speedy as it seems to be. I think if young people in the course of the Freshers' Dinner have two, or maybe three, glasses of wine, then this is ample and I see no need to go beyond this. Indeed I suspect that if water were provided, with the jugs and glasses on the table before dinner is served, the demand for wine expressed through glasses being empty would turn out to be rather lower. The addition to the task of laying the tables seems to be rather minor. The actual saving in wine and work of pouring it out may be significant.

This plea will strike a chord with those who were sitting on College Council in 2007 when the Dean and one of the Tutors expressed concern at the anti-social behaviour of certain Junior Members which was believed to be fuelled by the amount of alcohol being served at College dinners. As a consequence of the various discussions that followed, the 'two-glass rule' was imposed. Either Sir Hermann was ahead of his time or this is proof positive that things inevitably come full circle.

Science, Churchill & Me

The Autobiography of
Hermann Bondi

■

Master of Churchill College,
Cambridge

COMMUNICATING THE ART

I think you are a lousy professional if you know nothing outside your subject.[72]

As I started to archive Sir Hermann's papers towards the end of his Mastership I began to appreciate the breadth and depth of his knowledge and interests. As I filed away the documents I could see for myself that Sir Hermann seemed to have something wise to say about everything. He had already made known his view that everyone should be helped to develop their communication skills, saying that a mathematician who is thought to be brilliant but is actually unable to convey his brilliance to others is nothing more than a 'pain in the neck' and certainly not a contributor to mathematics[73] In his autobiography, Sir Hermann referred to some advice he would give when Secretary of the Royal Astronomical Society and a young person was due to give a paper:

> "You are addressing some very distinguished astronomers. Please speak to them as you would to twelve-year-old children." Only the few who followed my advice stimulated a good discussion".[74]

I recall a moment when Sir Hermann stopped what he was doing to talk to me about a book he had written. I cannot remember which one it was, but it was most definitely scientific and though I was a lowly first year Open University Arts student at the time, I had nonetheless felt compelled to congratulate him on the clarity of the first chapter and my surprise at believing myself able to understand what he was trying to convey. I can see him now, standing in the doorway between our two offices in the Master's Lodge, saying: "Well, I can see you clearly understood that bit, so let me continue a little further …" at which point my pride at having being thought by so great a man to have understood something so far above my own level of understanding intervened and the tenuous link between his world and mine was severed. He must have seen this, of course, but not for one moment did he let me feel inadequate in terms of the discussion in hand.

Yet he did not always suffer fools so gladly. There is a letter on file from a retired scientist who was claiming to have discovered flaws in Sir Hermann's explanation of the route dependence of time in his book 'Relativity and Common Sense'. A copy of the typed response is clipped to the original. It is dated 14 February 1990 and quite simply says 'I do not agree with you'. Another letter from a lawyer believing himself to have found the solution to an important problem of theoretical physics and managing to get referred to Sir Hermann by the Assistant to the late Sir Karl Popper, was left in no doubt as to Sir Hermann's assessment of his capabilities. His letter, dated 20 September 1995 says:

> Thank you for sending me your book. As for your 'deceleration' hypothesis, I can see no scientific significance in it.[75]

While so irritated was he by a request from the Editor of an International Biographical dictionary, that he instructed his secretary to reply on his behalf suggesting that he should 'first of all consult the current edition of Who's Who and then only ask necessary additional questions'.[76]

On the other hand, Sir Hermann could give the impression of having all the time in the world to deal with smaller requests, being happy to provide a signed photograph to add to someone's private collection of photographs of those people whose work and activities had made a 'notable contribution to the twentieth century', or to make himself available to a BTEC HND photography student from Bournemouth & Poole College of Art and Design who wrote in June 1989:

> I am currently studying photography at the above college and am starting a personal project on portraits of Cambridge University masters. My ideas for the photographs are to portray Cambridge masters with sympathy for their views on where and how they would like to be photographed. Very often I find that not enough consideration is given to the actual person concerned.[77]

He would also take time to respond to special requests such as one from a pupil at Archway School in January 1989. The pupil was one of a group of 5th and 6th year pupils engaged in collecting memories of the outbreak of the Second World War, the school having felt that the approaching 50th anniversary of the declaration of the Second World War would be an appropriate opportunity to learn more through the experiences of the people actually involved. Sir Hermann's generosity of response is typical. On 23 February 1989 he wrote:

> In the summer of 1939 I had a good and lengthy holiday in the Swiss Alps. I returned to my parents, who then lived in London, just before the pact between Germany and the USSR was signed, on 22nd August. Once one heard of this, the seriousness of the situation was plain to all. I felt that the best place for me to be was in my College (Trinity) in Cambridge and so I went there, on, I think, 1st September, just as the evacuation of children started. To my great pleasure I found that many of my friends had had the same idea, although it was over a month before our third year at University was due to start. I distinctly remember listening with perhaps four or five of my friends to the Prime Minister's broadcast announcing that the country was at war. We were all only too conscious of the dangers this portended for us but nobody doubted that the declaration of war had been unavoidable.[78]

Despite his exceptionally busy schedule, Sir Herman would often go out of his way to accept engagements he felt were worthy and often at some inconvenience to himself. For example, there is a delightful exchange of letters between Sir Hermann and the Head of Complementary Studies at Sir John Deane's College, in Northwich, Cheshire, who wanted to see if Sir Hermann would be willing to give a talk to her sixth formers.[79] Sir Hermann wrote on 10 March 1989:

> Certainly the idea of coming to your College and giving a talk to your Sixth Form appeals to me. However, I am a little worried by the practicability. I do not like driving long distances and prefer not to miss an evening here. I do not object to early starts and could take the 06.00 train from Cambridge, due at Nuneaton at 8.24, but the onward connections are poor. There is a flight from Stansted to Manchester, but how stable is the time table? Have you any suggestions?

The Head of Complementary Studies responded on 14 March:

> I was really delighted to receive your letter agreeing in principle to come to the College to talk to the students. As a consequence I've been doing a bit of research!

> You can get here and back in one day fairly easily if you don't mind early starts. There are two routes if you travel by train … In either case we can meet you at Crewe. The return journey also offers two alternatives …

Professor Sir Hermann Bondi, KCB, FRS

CHURCHILL COLLEGE
CAMBRIDGE
CB3 0DS, UK   **Fax: (0223) 336180**

Sir Hermann's business card

…Returning via London has the advantage that you would have plenty of time for lunch with us before you leave. It would also mean only one change of train. Perhaps you could come via Nuneaton and return via London. If none of these seems suitable then I will investigate flights from Stansted.

I have enclosed a list of available dates although I know it is too early for you to commit yourself to any of them yet.

At Sir Hermann's request, the correspondence was then placed in the Master's office 'bring forward' basket for review in September 1989. However, on 29 June, the Head of Complementary Studies wrote again:

Earlier this year you agreed that you would come to talk to the students at this College some time within the next academic year. Now that 'next academic year' is almost upon us I am trying to organise the speakers and dates for the session. I do realise how difficult it is to commit yourself to a date so far in advance; but I would be grateful if you could give me a date – however tentatively – so that I do not book someone else for the one day in the year you could have come! The dates available are [thirteen dates offered.]

The file shows that I was then asked to respond to this letter on 14 July. I wrote:

Thank you for your letter of 29th June. Although Sir Hermann is unable to offer you a firm date at the moment, there are two dates from the list you offer on which he may be free: 25th January 1990 and 2nd March 1990. We should be in a better position to give a firmer date at the beginning of next Term.

It appears that true to my word, I followed this up on 15 September:

The Master's private and official headed paper

Unfortunately, Sir Hermann now finds that he will not be able to manage 2 March 1990. He has an engagement on the previous day and will not be able to leave early enough the following morning to make the necessary train connections. It is very unfortunate that we have not been able to agree a date so far. You may find it helpful to know that Sir Hermann retires as Master of Churchill College at the end of July 1990, so it should prove a lot easier to find a convenient date from September 1990 onwards.

Sir Hermann then wrote on 27 September:

I am sorry it is proving so difficult to find a date. Of those you mention only 16 March and 27 April are possible and those two only with rather different timing, as on both occasions it is essential that I am back here by 5.30 p.m. Leaving a reasonable safety margin, this implies (assuming next week's timetable changes are not relevant) leaving Stockport at 12.25 p.m. Of course this would mean that I have to arrive the previous night, but I trust somebody can offer me a bed.

Please let me know how this would fit in with you. 27 April might be marginally easier than 16 March, but the difference is not great.

The Head of Complementary Studies responded on 5 October in somewhat less enthusiastic tones:

I too am sorry that it is proving so difficult to find a suitable date. Because we are trying to juggle with 700 students – 300 of whom will come to your lecture – it just isn't possible to alter timetables. Under our present system it just isn't possible to fit the lecture in to allow you to get to Stockport by 12.25 p.m. (staying the night is no problem. We'd be delighted to find you a bed!)

Under the circumstances it might be better to leave it till September/October 1990. Your secretary tells me that your schedule will be less circumscribed after the Summer so perhaps it will be easier to find a suitable time. We are very disappointed that we'll have to wait till then but perhaps it would be easier. Is it all right, then, if I write to you in April 1990 to make arrangements for the following academic session?

Thank you again for agreeing to come — and for trying so hard to find a suitable date.

Whether a date for the talk was ever agreed, the file does not reveal, but all credit is due to the Head of Complementary Studies for her perseverance.

However, as befits any good communicator Sir Hermann would sometimes offer advice to others. For example, on 27 April 1990 he wrote to the editor of a new journal:

You asked a number of questions on fees to be paid….. Basically, my outlook is that I am very happy to do useful jobs for nothing. Of course, if somebody chooses to pay me many hundreds of pounds for a job I would do anyway out of interest, it is agreeable, but to be offered £50 or the like seems to me almost an insult. A very high sum indeed would be required for me to undertake a job that I would not otherwise want to do. On the other hand, I regard the prompt and generous settlement of expense claims as an important element of courtesy. My advice, therefore, would be not to offer editors, reviewers or contributors any direct payment, but encourage them to send in their expense claims, promising that they will be dealt with speedily and generously.

… A little flattery never comes amiss, particularly if it is true. To write to one of the major contributors to the subject and say that nothing would start off the Journal as successfully as a review article by that person on a relevant topic is both true and attractive to the author.

….To make people travel (to meetings) is not only expensive but people of this calibre would not wish to travel unless the meeting has real point, important discussions and the like. A meeting at which everybody only exchanges pleasant platitudes will not endear the journal to its editors. [80]

While on 21 October 1987 he wrote to the Vice-Chairman of The Queen's English Society:

I was very much impressed by the letter you wrote to me and I fully agree with virtually all you say there. Like you, I feel that a better ability to write comprehensible English is an essential aim of school education.

Therefore I deeply regret that you have chosen to word the petition in such a specific and rigid way that I cannot see my way to signing it. While wholly applauding what you want to achieve and while entirely agreeing that the dropping of the teaching of formal grammar was a major mistake, I do not think it right to prescribe so precisely how teachers should achieve our common aim. [81]

Sir Hermann was a prolific letter-writer. Sometimes the letters were for public display: for example, in the twelve months from June 1988, he had eleven letters published in various newspapers on subjects ranging from student finance, the nation's health, through to the bankruptcy of communism. At other times the letters were to communicate an error, such as one Sir Hermann discovered on a visit to Peterborough Cathedral. On 24 October 1988 he wrote to the Dean as follows:

On a most enjoyable visit to your beautiful cathedral I went to look at the excellent display of your excellent clocks. I was horrified to discover that in this description there and repeated in the printed pamphlet, Gallileo is described as a famous astrologer. This is a disgraceful travesty of the true history for Galileo was a rational scientist, rightly famed for his contributions to physics and to astronomy. To describe him as a famous astrologer is an insult to his memory and propaganda for astrology, a superstition on which, so I understand, the Church frowns.

I trust, Sir, that you will ensure that this ridiculous mistake is immediately corrected and Gallileo is described as a famous astronomer, both on the notice by the clocks and in the pamphlet.[82]

From time to time he would send a letter to report a problem, such as in this letter to the Highways Department of Essex County Council dated 27 January 1989:

The footpath from Hadstock to Ashdon via Bowser's Farm is made almost impassable by the collapse of the footbridge between Bowser's Farm and Ashdon (O.S. Map 154, Grid reference 576 424). I trust this will be reinstated shortly.[83]

Or he would write a letter of complaint, such as on 7 August 1989 when he wrote to the Dean of Ely Cathedral:

In the afternoon of 31 July we paid an unpremeditated visit to Ely Cathedral with two elderly American relatives and the occasion was so unsatisfactory that I feel I must write this somewhat sour letter to you.

I entirely accept and support the charging of entrance fees to visitors, but surely this imposes an obligation on the management to make it absolutely plain what the tourist can expect for his money, an obligation that is in no way discharged by the general notice about the repair work, but would be discharged, for example, by a blackboard at the desk stating the day's constraints. What we did not know when we entered about 3.30 p.m. and paid was:

(i)   That on that day the Stained Glass Museum would close at 4.00 instead of 4.30 p.m. I understand this was due to an illness preventing the usual attendant being present but should one not learn of this on entering? We might well have hurried there and still got a glimpse instead of arriving only at 3.50 and finding it closing. (Incidentally, from my small statistic, the regularity of the Stained Glass Museum keeping to its hours is very unsatisfactory.)

(ii) That the Lady Chapel was closed.

(iii) That the lighting on this dull afternoon, particularly east of the transept and for the First World War Memorial would be so minimal as to diminish the pleasure of the visit materially.

I may say that when we mentioned this at the front desk as we were about to leave, the staff there entirely agreed and rang the verger to ask for more light, a request turned down with no excuse or reason given. Of course there may well have been reasons, perhaps connected with the building work, why the electricity demand had to be kept low, but one might have been told. Moreover, insult was added to injury by the blaze of lights in the shop…[84]

Or this letter that he sent to British Rail on 14 October 1988 after having arrived forty minutes late for an engagement in London:

On Sunday 2nd October, I had an important engagement in London. Following advice in your timetable about weekend travel, I rang Cambridge Station on the evening of 1st October to enquire whether on the morning of 2nd October normal Sunday service would operate between Cambridge and King's Cross and was told this was the case.

I therefore decided to take the 09.25 from Cambridge due at King's Cross at 10.40. After I had taken my ticket at Cambridge about 09.00 I was told that I would have to change at Royston. It transpired that there was no electric service to Royston and we were taken by diesel instead. The train was slow because of work in progress. At Royston I joined an electric train which then proceeded by Hertford North, the main line, so I was informed, being disrupted at Welwyn. The train stopped frequently and I arrived at King's Cross at 11.20, making me forty minutes late.[85]

But this is far from the ranting of an inveterate or irascible complainer. I know because I transcribed many of these letters from my own shorthand notes. I can still hear Sir Hermann's only very slightly irritated but confident voice articulating the problems in a clear matter-of fact manner. These were matters that affected other people and needed to be communicated. That was his primary intention.

Whether through his academic research or other interests, it can be seen in the documents, letters and other papers in the files that Sir Hermann worked tirelessly to communicate his art, whether in writing or orally. With respect to lecturing he once wrote:

I greatly enjoyed lecturing, which forced me to think matters through until they were so very clear to me that I could lecture without notes. This had the great advantage that I could watch the class and observe the warning looks of utter incomprehension![86]

But with respect to his role as a manager he wrote in his autobiography:

Almost every problem that an organization suffers from arises from insufficient communication. For a compulsive talker like me, this was a relief to realize. I appreciated how much I could do, simply by going round talking to people and listening to them. Almost every failure in management is a failure in communication. Of course, decisions have to be taken, but decisions arrived at after ample communication are not only better decisions, they are much more widely accepted and understood.[87]

However, there was one area in which his communication skills did not always achieve the desired results. The former Bursar, Hywel George, recalls:

Sir Hermann accepted that fundraising was part of the Master's role but was not at his happiest in face to face meetings with potential donors. I did not accompany him on many such visits but I recall one meeting with a leading industrialist. After considerable trouble we had succeeded in having a 30 minute appointment in London. For the first 25 minutes or so the industrialist listened to Sir Hermann holding forth on the relationship between science and technology but without a mention of the College's specific needs and why the meeting had been arranged. When I saw the industrialist looking at his watch I had to remind Sir Hermann of the purpose of the meeting but by then it was too late and we came away empty handed. On the other hand, armed with a collecting tin outside a Cambridge store or railway station, he was the Save the Children's star collector.

That Sir Hermann was able to convey his knowledge and interests to other people in such effective ways is beyond doubt. That he was able to do so with people of all levels of understanding is to me the point where his brilliance shines through.

View across the playing fields towards
the Chapel at Churchill College and the
Sheppard Flats (1984)

OF HUMAN ENDEAVOUR AND CO-OPERATION

Science, Sir, is about bringing knowledge out of the unknown, through uncertainty, through hypotheses and through interpretations. Dogmatic answers are only seen by those who crave for them.[88]

As noted in the Preface, this memoir makes no attempt to applaud Sir Hermann's scientific achievements, offer intellectual debate on his views on education or comment on his stance on religion. Such a task has already been undertaken by those who are qualified to do so and I am clearly not such a person. What I am seeking to convey, however, is a flavour of what was happening below the public image, the day-to-day and the mundane; in short what it was that made him tick.

It will probably come as no surprise to hear that Sir Hermann's College room contained file after file of articles, correspondence, papers and books on scientific, educational and religious – or humanistic - matters. Many of these files were sifted through on his retirement as Master and transferred to the Archives Centre, but many he chose to transfer to his College room. One such file contained a full page article which appeared in the Times Education Supplement in 1982 shortly after news of his appointment as Master. It was entitled 'Why science must go under the microscope'. In the article Sir Hermann argued that the education system was offering young people 'a false science prospectus'. He believed that too much emphasis was being placed on the results of scientific research and not enough on the acquisition of problem-solving and communication skills. Never one to be shy of controversy he asserted:

> What we convey before the age of 16, as well as after it, is so false a prospectus of the subject that were we in business we would be accused of fraud. [89]

Alison Finch, Fellow and Vice-Master at the time of writing, recalls a comment he made on the teaching of science in schools that she says she has often thought about subsequently:

> He bemoaned the fact that this almost invariably concentrates on the verifiable, on certainties; whereas in Hermann's view science could be made much more exciting to schoolchildren if what was stressed was its uncertainties.

Sir Hermann felt strongly that science should be made much more attractive to much larger numbers of people:

> …. by teaching it as it really is, as a human endeavour that requires cooperation and communication, a human endeavour in which the efforts and steps that turn out to be unsuccessful enormously outnumber those that are successful. We will then have in our society more people who appreciate the nature of science: people who realize what kind of questions you can ask in science and what kind of answers you can expect.[90]

RSA

HIGHER EDUCATION FOR CAPABILITY

update

ISSUE 2 ■ FEBRUARY 1990

HEFC Update is the newsletter of the Higher Education for Capability initiative. It provides:
- news of activities
- a forum for debate
- examples of good practice
- reports of developments, and
- information on books and resources.

READERS: PLEASE WRITE!
HEFC Update is sent to every higher education institution, to a wide range of employers, to validating and professional bodies, and others interested in HEFC. The readership in higher education consists of people at all levels, including institutional managers, course designers, teachers, and students.

Readers' contributions are welcomed. Limited copies of back issues are also available upon request. For details of how to receive individual or multiple copies, please write to the address on the back page.

Higher Education for Capability is an initiative of the RSA (Royal Society for the encouragement of Arts, Manufactures and Commerce). The RSA sustains a forum for people from all walks of life to come together to address issues, shape new ideas and stimulate action. The Society works through lectures, seminars, conferences, projects and awards. Independent of special interests, it has been a seed-bed for new ideas and an agent for change since its foundation in 1754. More than 15,000 Fellows (FRSA) world-wide support the RSA's work.

Education for Capability is described in full on the back page.

Reflections on Higher Education for Capability

by Sir Hermann Bondi

I am delighted to be involved with this project. The coming decade will present those concerned with higher education in the UK with many difficult choices. I trust the project to generate results that will enable better choices to be made.

It is clear that higher education is closely related to all other parts of the educational system, and not only because we receive our students from it and it relies on graduates to do much of the teaching and management. Degrees, rightly, have their place among all the available qualifications. This country has a special problem because too few non-degree qualifications enjoy the esteem that they have in many continental countries. Until this is remedied, people will come to study for degrees not only because they want to do so (the only sound reason) but also because no other qualification is thought by them to be useful. Therefore the future of higher education is tied up with the success of the whole educational system and the reputation of all its certificates. The RSA's Education for Capability efforts (which I wholly support) contribute to improving this situation.

The success of higher education will be seen by our graduates as enabling them to lead fuller and more satisfying lives. Often this will involve a successful career and challenging jobs.

We are all convinced that this country needs more and better graduates. To provide them we need to address three constituencies:

(i) prospective students of all ages need to appreciate that higher education is **for them** and will be worth the effort and time. Clearly the presently under-represented groups should be addressed with particular attention, viz. women, some ethnic minorities, lower socio-economic strata, and people past the usual age for entry

(ii) students in higher education: They need to be involved deeply and directly in the learning process, not just through the prospect of exams. They must be enabled to help their teachers to be of the maximum benefit to them. They need to learn not just their chosen subject, but the skills of communication and of teamwork so essential to an effective life in any field

(iii) the prospective partners, colleagues and employers of today's students have an essential input to make into the whole scheme of higher education. The more directly and clearly this input can be articulated the better. We must learn what industry and commerce want, not what government **thinks** industry and commerce want.

One particular future employer is ready to hand and has a special input to make, namely the academic community itself. If the output of universities and polytechnics were unhappily to disappoint, say, an industrial sector, they could get their future staff from elsewhere, but higher education alone can provide the academic succession. If we were to lower our standards of excellence in this respect, we would let down the future.

Vital as this is, the more we can hear from all others what they find most valuable in degree holders, the more we can work to integrate their input into the system.

■ Sir Hermann Bondi is Master of Churchill College, Cambridge and has agreed to become a member of the Higher Education for Capability Steering Group.

However, he took every opportunity to ensure his views were promulgated through practical action at all levels, not just the high profile. For example, on 7 January 1987 he wrote to the Chairman of a small company just outside Cambridge:

> Over dinner at last night's excellent meeting we talked about science and mathematics education. I enclose reprints of mine on this, as well as the summary of my wife's "Girls and Mathematics" report.
>
> None of my writings, I am afraid, fully represent the strength of my anti-A level feelings.[91]

In late 1989 Sir Hermann and Lady Bondi published jointly in SCOPE, a magazine published quarterly for the youth section of the British Association for the Advancement of Science. Their article demonstrated how mathematics could be used to solve some very down-to-earth problems, such as the 'everyday frustrations' of the traffic jam.[92] Although these are just small examples of Sir Hermann's views on science, they do illustrate the breadth and sincerity of his attempts to convey his views to a wider public.

It was inevitable that Sir Hermann would take every opportunity to promulgate his views on educational matters. On 23 August 1982 a former Senior Tutor, Dick Tizard, wrote to Sir Hermann:

> If there is any way in which I can help by giving you information about the College I will be very pleased to do so. You will no doubt wish to consult primarily the principal officers of the College, but I might be able to add something since I am a founding Fellow and was Senior Tutor during the period when we adopted the policy of encouraging admissions from state schools, which I was so pleased to hear, in our discussions "in committee", that you evidently approved.[93]

Sir Hermann had responded a few days later saying that he was 'most anxious and keen to support and maintain the policy (which has been so successful) of encouraging admissions from state schools'.[94] And there is evidence in the files to show that he did indeed support the principle of widening participation and the need to ensure equality of opportunity. For example, Minute 9975 of the meeting of College Council on 14 February 1984 regarding Undergraduate Admissions, notes that the Master 'suggested that the comparative table should in future show the percentage of women applying to read arts and science subjects.' But Sir Hermann did not just exercise his skill at rhetoric, he also took practical action. Two examples of such action follow.

Cambridge Executive Education Programme

Among the files remaining in Sir Hermann's College room on his death was a file entitled the 'Cambridge Executive Education Programme'. Somewhat unsurprisingly, Sir Hermann's views were broadly-based and his educational concerns did not focus solely on primary and secondary education alone; he also felt there was a need for reform in the tertiary sector. The Cambridge Executive Education Programme was an attempt to address the so-called 'Cambridge Management Gap' of the 1980s. This was an initiative to promote the growth and better management of the small, new, high-tech companies growing up around the University with a view to helping young managers to understand better the elements of management and entrepreneurship. The intention was to create a programme for successful outward-looking executives with five years' experience. Sponsored initially by a small number of colleges, it was intended that this should make a major contribution to management education through a distance learning programme of courses and seminars.[95]

The Churchill Management Studies Group was thus established with the meetings being held in the Cockcroft Room at Churchill. I had the particular pleasure of minuting many of those meetings and was to experience for the first time the excitement of being part of an important initiative, particularly when Professor Charles Handy, a major founder of the London Business School and a world-renowned management thinker, was co-opted onto the Group. He was to prepare a key paper for the Group entitled The Education of Managers at Cambridge which still sits in Sir Hermann's files.[96]

In a letter to potential industrial partners in 1989, Sir Hermann wrote:

> From the beginning we felt it was most important for such a course to be both flexible and highly responsive to the needs of business, as business perceives them. The flexibility requires us to stay out of the University's system since this can only give qualifications on the basis of fixed, firm printed regulations. Thus the scheme will run under the aegis of several colleges without a formal University qualification on completion.[97]

The Group managed to capture the interest of five industrial partners and continued to develop its plans over the next five years until 1990 when the newly created Institute of Management Studies declared itself unable to take responsibility for the Cambridge Executive Education programme. It gave as its reason for this withdrawal of support the need to focus efforts on getting the MBA and MPhil programmes up and running. Although perhaps understandable, this was a major blow to the initiative and the final document in the file contains a letter from Sir Hermann dated 11 June 1990 to the remaining members of the Group:

> Because I feared such an attitude by the University, originally I wanted the whole issue totally separate from any University scheme. My difficulty in recommending that it should be carried forward as a purely College initiative is that I am not clear how we could negotiate ability to attend University lectures for the participants in our course.[98]

Although this initiative was not to see the light of day, it does demonstrate the strength of Sir Hermann's commitment to his cause.

Higher Education for Capability

But even as the lights began to go out over the Cambridge Executive Education Programme, Sir Hermann had already started to focus his attentions elsewhere.

> 'Higher Education for Capability' was an RSA project which aimed to encourage higher education institutions to give more students more opportunities to develop capability through the actual processes by which they learn. Specifically, it aimed to encourage students to be responsible and accountable for part or all of their own learning, both as individuals and in association with others.[99]

Sir Hermann's particular interest in this area is summed up in a letter to Professor John Stephenson, national director of the Higher Education for Capability initiative:

> I retire from this position in summer 1990. However, I will not abandon my relevant interests, which I might describe as follows:

> (i) widening of access to Higher Education, particularly as regards age range (mature entrants) and as regards those who have only received indifferent schooling

(ii) avoidance of over teaching, i.e. learner responsibility

(iii) the teaching of science and mathematics not as a professional preparation for future scientists and mathematicians, but in order to widen the general capability of those going into management, finance, administration, etc.

(iv) strengthening of links with continental Europe

(v) education for responsible citizenship ...[100]

Sir Hermann had some initial discussions about the initiative within Cambridge and in an exchange of correspondence with Dr Tess Adkins, then Senior Tutor of King's College, he asserted:

> What I think does apply here (to Cambridge) is that in some subjects, generally on the science side, there are more lectures than is ideal from an educational point of view. A little pressure to reduce lecturing (without reducing supervisions) might be useful, particularly if the proportion of mature students increases.

> ... I quite agree that there are a few individuals who are creative and intellectually able and not only are not 'rounded' but resist being pushed in that direction. However, there are fewer who deserve to be protected from being 'rounded' than think themselves to be in that class and, secondly, everyone should be helped to develop communication skills. The mathematician who, it is thought, is brilliant but is wholly unable to convey to other mathematicians what the brilliance consists of, I regard as a pain in the neck and not as a contributor to mathematics.[101]

Persuaded of the value of this initiative, Sir Hermann was to become a member of the Higher Education for Capability Steering Committee and to speak at conferences and other events in its support. In his article 'Reflections on Higher Education for Capability' he wrote in February 1990:

> They need to learn not just their chosen subject, but the skills of communication and of teamwork so essential to an effective life in any field ...

> It is clear that higher education is closely related to all other parts of the educational system and not only because we receive our students from it and it relies on graduates to do much of the teaching and management. Degrees, rightly, have their place among all the available qualifications. This country has a special problem because too few non-degree qualifications enjoy the esteem that they have in many continental countries. Until this is remedied, people will come to study for degrees not only because they want to do so (the only sound reason) but also because no other qualification is thought by them to be useful. Therefore the future of higher education is tied up with the success of the whole educational system and the reputation of all its certificates.[102]

Sir Hermann's support for the initiative, which ran for some ten years between 1988 and 1998, was greatly valued by Professor Stephenson who says:

> His support was key in helping us persuade the traditional university scientific establishment to take seriously the notion that the development of personal capability was as important a function of higher education as the acquisition of scientific expertise and that the former could be developed by the way the latter was taught.

Professor Stephenson went on to compile the HEC archive which is used today as a web-based resource for present day innovators and researchers into the debates of the time.[103]

Humanism

Of course, no memoir of Sir Hermann, at whatever level, would be complete if it did not mention humanism and this is obviously no exception.

Sir Hermann was President of the British Humanist Association from 1982 – 1999. In an interview published in *Science Age* in 1986, he was asked whether he agreed with the popular view that science and religion were complementary:

> I am a non-religious person … I see a very considerable difference of approach (in) that science is eternally provisional, eternally liable to be disproved, always concerned with things that may turn out to be otherwise. Religions depend on revelations that are supposed to be something unalterable, fixed and that they are for ever. That I find difficult to swallow. I very often do not describe myself as an atheist or humanist or agnostic but as an anti-revelationist. I strongly object to those people who claim they have a private telephone line to the office of the Almighty. That I find most repulsive. And when people, on the grounds of the supposed revelations, display great cruelty towards other people, try to tell non-believers what to do and what not to do, then I get very very bugged.[104]

Writing as President of the British Humanist Association, he articulated his views on secularity in the schoolroom in a letter to the *Times Education Supplement*, dated 21 November 1988:

> In your issue of 18th November two letters discuss constructively how the potentially divisive consequences of the religious clauses of the Education Reform Act can be mitigated. As a Humanist, I warmly welcome proposals such as alternative secular assemblies and activities with a strong moral ethos that will bring children together rather than separate them. Sensitive and thoughtful measures can avoid giving parents the awful choice between singling out their children by withdrawing them from assembly and R.E., or allowing them to take part, thus setting up conflicts with the ethos of their home.[105]

The College had no argument with his views or indeed with his public face. Fellow and Vice-Master at the time of writing, Alison Finch, writes:

> His public face not only as a distinguished scientist but also as a leading representative of British secularism also seemed to many of us to exemplify the spirit in which Churchill College had been founded.

Although Sir Hermann had no hesitation in making his views known, he did not let them affect his interactions with others at a human level. Another Fellow, Dr Ken Livesley, recalls:

> The 21st anniversary of the opening of the chapel took place during Hermann's Mastership. To celebrate the occasion Bryan Spinks, who was Chaplain at the time, invited a number of distinguished preachers, one of whom was the Archbishop of Canterbury, Robert Runcie.
>
> As Chairman of the Chapel Trustees I felt that Hermann should at least be formally notified of Runcie's visit, so I wrote to him asking if he wished to be involved in any way. He replied with typical Hermann

warmth, saying that he hoped I would understand if he didn't attend the service, but that he would be delighted to preside afterwards at dinner.

It turned out that they had met previously on a number of occasions and I think that Christine and Rosalind also knew each other. I remember it as one of the most enjoyable high tables I can recall, with conversation on a wide range of topics and with no sense of controversial subjects being either sought or avoided. Hermann even went out of his way afterwards to say how much he had enjoyed the evening.

I do not remember any feeling of surprise that two people holding such different beliefs should get on so well – it was, after all, only what one would expect from Hermann. But on reflection I think it does say something about the College as a place where such a meeting could not only take place but seem entirely natural.

I was to experience this also when in September 1989 my husband died suddenly, leaving me with two young sons to support. Sir Hermann was supportive and helpful, but never dogmatic and allowed me to come to terms with my loss in my own way. As he was President of the British Humanist Association at the time I had access to plenty of humanist literature and although he answered my questions with clarity, patience and empathy, I never for one moment felt any pressure to make my own views known or to follow a particular course of action. To this day I believe that this was a real endorsement of his great humanism.

I can recall a further example of Sir Hermann's humanism during the late 1980s when he was on the Editorial Board of the *Longman Encyclopaedia*, as well as being a contributor to the physical sciences section.[106] I recall an occasion when he asked me to "ring up the European Space Agency" to check some details on, I think, the space rocket Ariane. I was not too convinced that the European Space Agency would be that willing to reveal the information to me, but I was careful to let them know that I was ringing on his behalf and after a few transferred calls within the ESA it seemed to do the trick. I duly returned to Sir Hermann with the required information. I am still slightly alarmed at the confidence he placed in me to obtain the correct information for so important a publication, but Sir Hermann clearly felt that I could do this and was content to place his trust in me. Years later I can recognise that these were his humanistic qualities shining through: he was endowing me with individual responsibility and respect.

Naval Radar Reunion
1st April 1995
to be held in
Churchill College, Cambridge at 6.00 pm

This is not an April fool, far from it but notice of a reunion to celebrate the completion of 10 years work by many members of the Trust and the publication of the two companion volumes to "Radar at Sea".

The programme will be as follows:

6.00 pm	assemble in the Jock Colville Hall
6.30 pm	short talk by the Archivist, Alan Kusai, introducing the Archive Centre
7.00 pm	short tour of the Archive Centre
7.30 pm	dinner in the Fellows' Dining Room presided over by Sir Hermann Bondi KCB, FRS.
10.30 pm	carriages

The Trustees have already invited the First Sea Lord, Admiral Sir Benjamin Bathurst, GCB, ADC and his lady and will be inviting a number of other distinguished guests including Admiral of the Fleet Lord Lewin, who wrote the foreward to "Radar at Sea", Admiral Sir Jeremy Black, who wrote the Epilogue, Admiral of the Fleet, Sir William Staveley, who represented the First Sea Lord at the 1990 reunion, with, of course, their ladies. They will also invite the present Master of Churchill College, Professor A.N. Broers, F.Eng FRS, the President of the Royal Society, Sir Michael Atiyah OM, the President of the Royal Academy of Engineering, Sir William Barlow, F.Eng, all of whom have contributed generously to the project, with their ladies.

Since this will probably be our last reunion it is hoped that everyone who can come will do so. Arrangements have been made so that those who want to spend a little more time in Cambridge can have lunch in Churchill College on either the Saturday or the Sunday or both. Tea will be available in Chuchill from 4.00 pm onwards on Saturday. If you wish to come will you please complete the enclosed form and return it to me at the above address by 10th March 1995 together with a cheque made out in my favour.

I do not expect that there will be any shortage of places but please book early to ensure that there is room for you. If for any reason you are unable to reply before 10th March do not hesitate to 'phone me and I will then do my best to fit you in.

John Coales

Cambridge
12th January 1995

TWO FLYING ROAST DUCKS

Man who waits for roast duck to fly into mouth must wait very, very long time.[107]

In the Preface to his autobiography Sir Hermann said:

> Sometimes it seems to me that I have been walking through life with a wide open mouth and roast ducks have come flying in with monotonous regularity …I have never had a particularly high opinion of my intelligence and have told my children, only half in jest, that it was much more important to be lucky than intelligent. They are quite ready to believe that I am lucky![108]

That he was lucky enough to enjoy robust good health and to have boundless energy is beyond dispute. He had a prodigious output and a check of his diary and logged activities for June 1988 alone reveals the pace of his life. For example, at the beginning of the month he had chaired a meeting of the College Council, dealing with issues from anti-social behaviour (and in particular 'the practice of throwing people into the ponds'), the updating of the heating system in the Wolfson Flats with electric storage heaters and the computerisation of the Archives Centre.[109] As chair of the Colleges' Committee he was leading the colleges on strategic issues such as the Education Reform Bill, public relations and Project Granta (the introduction of computing into the University and its extension to the colleges); he had just returned from a conference in Bologna where he had spoken on the History of Modern Cosmology; he was preparing for an International Science Policy Foundation event on 8 June and was due to visit Maastricht for several days on behalf of IFIAS (International Federation of Institutes for Advanced Study). He had also started work on his autobiography, commented on a paper by Sir Harry Hinsley on climatology and on a report of the Royal Commission on the NHS and had three letters published in national newspapers. In addition to this he would have been running his office, hosting the usual end-of-term social events and dealing with the other College business that inevitably emerged as the academic year neared its end.

Sir Hermann once described a day in his life in early 1988:

> Only recently, just before I was sixty-nine, I woke up early one morning in Toronto, took the breakfast flight to New York, saw my sister, took part in the Board Meeting of the Winston Churchill Foundation of the United States, went in pouring rain to the airport, flew home, arrived in the Master's Lodge a little after ten o'clock and within half an hour was chairing the meeting of the Heads of Colleges. That afternoon we had a party of students for tea and that evening a reception and dinner for new Fellows.[110]

With scant regard for how tired he might have felt – or at least how tired one would have expected him to have been feeling – he would have approached every appointment or event that day with the same degree of enthusiasm and commitment. The students at the tea party would not have been short-changed.

But if it was his boundless energy that enabled him to grasp at opportunities, his insatiable curiosity and inherent generosity were arguably the real keys to his success. If something captured his interest, he was committed, either in the role of co-author and mentor, or as the main driver of a project. Two such examples of these are detailed below.

Magic Squares

As I am sure any Master's Secretary in Cambridge will agree, one of the most exciting parts of the role is the contact with people whom one might otherwise never have met. A particular example of this for me was Dame Kathleen Ollerenshaw.

Dame Kathleen had a fascinating background. Although almost completely deaf from the age of eight, she studied mathematics, battled her way through to an Open Scholarship to Somerville College, Oxford, graduating in 1933 with a first class degree. In 1936 she started research on cotton for the Shirley Institute:

> She taught herself all the statistical techniques that she needed to test the efficiencies of the different methods and ingredients used in weaving. By applying advanced algebra, she managed to discover methods to complete a task in six hours that usually took six days. After this, she was offered a permanent position designing a cotton canvas that would be impervious to water for making tents for the army.[111]

She went on to work alongside Alan Türing at Manchester University, was appointed Dame Commander of the British Empire for services to Education, was elected Lord Mayor of Manchester in 1975 and became a Founder Fellow and later President of the Institute of Mathematics and its Applications, for which she wrote many papers, including the first general solution to the Rübik's cube.

In 1982 Dame Kathleen and Sir Hermann jointly published a paper entitled Magic Squares of Order Four and an original copy of this text was stored in Sir Hermann's College room in a yellow plastic folder.[112] Dame Kathleen was wonderfully eccentric and the correspondence relating to this work brings to life that sense of excitement that must accompany the certain knowledge that one's endeavours are about to bear fruit. On opening the yellow folder, one's attention is immediately drawn to a letter which has been carefully inserted at the front. It quickly becomes clear that Dame Kathleen was writing while on a train journey. The letter is headed 'In train: 8.12.81' and she writes:

> The miracle achieved…. I do hope you think it now just about as good as we can get it and as 'free from error' as one dare hope in a fallible world.

The text itself bears the acknowledgement:

> 'To Hermann … immeasurable thanks for all the marvellous fun – we've made it!'

Dame Kathleen's work on Magic Squares was to continue for some years after this, but she kept in close contact with Sir Hermann, whom she clearly regarded as a mentor. She was later to send him a green folder containing the Enumeration for all Most-Perfect Squares, a paper that was published by the Royal Society in 1988.[113] Her handwritten letters are quite unique in their style, where the thrill of discovery is juxtaposed against everyday events.[114] The following letter was dated 8 November 1986:

Dame Kathleen Ollerenshaw, DBE

Isn't this thrilling; received today… Progress here is immense – but totally time-consuming. I have put down a deposit for (don't have a fit) two! utterly adorable black Miniature (not Toy) poodles to take delivery on 27th Nov …… I want a pair of (male) 'mathematical' names. Any ideas? I thought of Gauss & Galois but too weighty for such little dogs. 'Plus and Minus' makes one inferior and Max & Min makes one feminine. We could have Archie for Archimedes but what about the twin brother?

It is fascinating how the incredible and the banal stand side by side. Dame Kathleen wrote again on 8 April 1987:

'Sick' with excitement - ½ hour ago – I must share with you prompto "& before I get smashed on a motorway so that there is a record". I sweated away and have multi-checked full computer prints for $12 \times 12 = 42$ and $24 \times 24 = 210$……"

[Mathematical formulae follow]

….The patterns for binary with only binary factors to consider at last emerged through the method & back at home I checked and refined my new shorthand notation. There had to be a formula.

It's supreme & I can prove it ...[More mathematical formulae]...Please, please say that this is absolutely stupendous....

A year later, she wrote to Sir Hermann again:

In Train, Stockport Watford junction, Bank Monday 30 May 88

....I'm aware that you are sure to comment that the explanation is (as yet) totally inadequate, i.e. just why these and only these patterns lead to a unique principal solution for each set ...

... Here is one of the bits of advice I'd welcome. How general to make the introduction and 'method explanation' and 'proofs' ...

Sir Hermann responded on 3 June 1988:

Yes, your splendid green folder arrived safely last Saturday morning and has given me a great deal of interest and fun.

There is no doubt that it is a massive effort revealing numerous remarkable features. What I found particularly intriguing was the uniqueness of the principal solution in the case of 'twists'.

Your method of presentation has all the characteristics I like, especially in being so concrete and specific. However, when you write up for publication you must aim at the pure mathematicians who prefer a more abstract form of presentation. There is also the worry about length. It is now considerable and making it more abstract is not likely to lead to a large reduction in length, though the amount of tabular material will no doubt be less.....

...Anyway, it is all most impressive, but the magnitude of the effort still needed to get it into a form suitable for publication is considerable ...

Dame Kathleen replied on 9 June 1988:

Yes, not only has there to be much clearer explanation/proof that the method gives all solutions and no repeats (as I knew to be necessary) and how it really works, but, as I also knew, it will have to be 'dressed up' ....Where I have been somewhat troubled in the prospect of this Big Task has been that it is in a relatively recently highly developing field of 'Combinatorials' (if I have this right) whereas I stopped at the 'Permutations & Combinations' of sixth form algebra all those years ago. I got out a book from UMIST Library and took it back a week-or-two later. All the symbolisms were so new to me, it might have been Chinese and I couldn't see that, even if I'd ploughed through to understanding, it was likely to make any significant contribution to my Squares.

Another (undated letter) is written from 'Here & There. Now at Hairdresser. Friday.' Whereas some women might be relaxing at the hairdresser, day dreaming perhaps or reading a magazine, Dame Kathleen was writing a six page hand-written letter to Sir Hermann, talking about the difficulties of getting used to word-processing, getting amended print-outs and 'saving' documents, finding a way of expressing the total number of lead sequences, or making the 'twists' more sophisticated. A postscript to the letter, which reveals that the dogs were eventually called Max and Min, updates Sir Hermann on their progress:

> Max had a patch of hepatitis – very sad little dog but modern science with four days' injections did the trick and he's marvellous again. Min just hung around and sniffed lovingly. Max went on chicken-only (& I had to let Min join) – it's taken a week to wean them back onto ordinary Pedigree Chum! (prefer chicken).

I remember one occasion when Dame Kathleen visited the College. Sir Hermann had invited her to one of the College Feasts as an official guest and I remember being introduced to her. It came as no surprise to discover that in real life she was just as eccentric and delightful as her letters portrayed. Garrulous and friendly, I don't believe she stopped talking for a moment. Christine reminds me that it was on that occasion that Sir Hermann received an SOS from the petrol station on Huntingdon Road. Dame Kathleen had driven down from Manchester and had stopped off to fill up with petrol before arrival at the College. It was only then that she realised that she had left her handbag in Manchester. So minutes before she was scheduled to attend the reception for Official Guests, Christine had to rush to the petrol station to bail her out. But all was resolved with great humour and hilarity on both sides.

Naval Radar

> These same Wise Men, it is related, when about to engage in any Venture, do make use of that which they name the Devil Finding Mirror, the same being a species of Speculum wherein are displayed such Demons of the Middle Air as may draw nigh to molest them on their Occasions.[115]

Sir Hermann's interests were of course very diverse and one project in which he became heavily involved was contained in several of the files remaining in his College room on his death. The files related to Naval Radar. In the late 1980s Sir Hermann had been closely involved in a project to assemble 'a comprehensive collection of archives relating to the history of British naval radar from its inception in 1935 to the end of World War 2, not only for the historical record, but also in the hope that one day it would lead to a published account.'[116] The idea had been the brainchild of Professor John Coales, Emeritus Professor of Engineering and a scientific civil servant prior to the start of the Second World War. Since those concerned were in their sixties, seventies or eighties, it was important that the data was gathered as soon as possible.

A reunion at Churchill College on 12 December 1986, which was attended by nearly fifty officers, both civilian and naval, who had served at the Admiralty Signal Establishment before 1946, was to see the creation of the Naval Radar Trust. Sir Hermann was elected Chairman. Derek Howse, former head of the Department of Navigation and Astronomy at the National Maritime Museum, was commissioned to write the book.

The papers show Sir Herman's great attention to detail. In a letter to John Coales dated 27 October 1988,[117] he writes:

> … I am not happy about the timetable: the first three dates are fine for Parts 1 and 2, but I think much of Part 3 can scarcely be produced until one has Parts 1 and 2 in one's hand, nor, I suspect, can Parts 4 and 5 be completed in the same time scale as 1 and 2. I would therefore suggest that Part 3 contributions can be received until 1st May …

In a letter to Basil Lythall, former Chief Scientist, Royal Navy, dated 16 December 1988[118], Sir Hermann comments on a write-up of a chapter entitled 'Basic Science and Research' which was intended to form the basis for a chapter in Part 2 of Derek Howse's book:

.... p. 5 para 4, line 7: 'not always up to full efficiency' is an extreme euphemism. If I rightly remember, to be only 10db down was pretty good.

p.6 end of para 2: I think this should be expanded. No modern audience appreciates how much cunning, skill, mathematical ingenuity and sheer hard work went into computing them.

p. 7, end of para 4: Not only was the railway commandeered but also the Café on top where I lived for many months, while Goodwin handled the Aberporth end. This might be mentioned explicitly, as also the difficulty of finding long over-sea ranges which led us to use Snowdon.

p. 8 footnote: Gold's work on high speed photography made it a tool in the laboratory, but perhaps the preceding pioneer efforts of Bolitho might be mentioned…[119]

It was clearly important to him that the facts were correct, but so was appropriate recognition of the achievements of others. For example, in the letter to John Coales cited above he also wrote:

…I joined ASE only on 1st April 1942, but the air was still humming with praise of Landale for having pushed 271 out to the Fleet at such speed and how ably he had short-circuited Shuttleworth's designers (always very slow) to achieve this. Surely this should be mentioned? History need not be kind.[120]

Sir Hermann was to attend countless meetings of the Naval Radar Trust and the papers contain many handwritten letters from John Coales, the co-ordinator of the many strands of the project. Sir Hermann's responses as Chairman would always be encouraging and to the point. For example, at a meeting of the Trust on 5 July 1988 when it was agreed in his absence to purchase a copy of 'War at Sea by Stephen Roskill for £250, in a letter to John Coales dated 2 September 1998[121] Sir Hermann enquires '…. Surely Derek Howse could borrow it and photocopy the pages concerned for far less than £250? Sorry to be fussy, but as Trustee, I ought to be inquisitive'. He is clearly satisfied with the explanation since he replies on 28 September 1998 'Thank you for your letter. With your explanation I am very happy to agree with the Minutes.' In another letter to John Coales dated 21 November 1988 the message is very succinct: 'Dear John, I am now wholly content.'[122]

The book, eventually entitled Radar at Sea, was published by Macmillan Press in 1993, with a book launch on 26 February 1993 at Imperial College of Science, Technology & Medicine, in London. But in addition to the book, it had been the stated aim of the Trustees to locate, as far as possible, all archival material relevant to the development of British naval radar and its operational use in World War II and ensure its safe keeping as far as possible[123]. An agreement was therefore drawn up between the Naval Radar Trust and the Master, Fellows and Scholars of Churchill College to hand over the whole archive and to hold it as their sole property in perpetuity. The archive was estimated to comprise well over 95% of the documentary evidence required to support a true account of the development of radar for the Royal Navy between 1935 and 1945.

At the final reunion dinner of the Trust held in Churchill College on 21 July 1990, Sir Hermann paid homage to naval radar:

In that desperate struggle the one area that really worried Winston Churchill was the Battle of the Atlantic. To keep open the Atlantic lifeline was enormously difficult and the dangers were tremendous.

Naval radar made a very major contribution to this but it is above all the gallantry, the skill, the determination of the Royal Navy that won that most crucial of battles. It is with the Royal Navy that we scientists and engineers worked in such fruitful, indeed decisively fruitful partnership …[124]

His secretary at the time, I typed that speech for him. I can see him now as he paced between his office and mine, speaking confidently, carefully articulating the words for my benefit and with an impassioned and uncontrived confidence in what he was trying to impart. My shorthand notepad would fill rapidly as I took the almost word perfect dictation. Such was his conviction and skill as an orator that I would be carried back with him, seeing imagined pictures of a cruel sea and of battleships in the Atlantic.

As Domestic & Conference Manager, Anne Hammerton was closely involved in the arrangements for the meetings and particularly for the celebratory dinner. The meetings were generally held in the Master's Lodge and would inevitably take place at weekends, with one or two of the participants sometimes staying overnight. Anne says she got to know them all rather well and as a consequence was invited to attend the final dinner as a special guest. She says that she was terrified, feeling she would be a fish out of water, but Sir Hermann and the other Trustees were insistent, so Anne accepted. She says that she has never enjoyed anything so much. She learned a great deal and the evening was a great success.

Following the reunion dinner, a letter from Captain Sir David Tibbits, DSC, FNI, RN, dated 24 July 1990,[125] acknowledged the important role that Sir Hermann played:

> …If the ultimate result is a successful history, we can all be pleased but we should not forget that the "whole" story – thanks to you – will be in the Archives of Churchill College.

There is a delightful postscript to this letter. Captain Tibbits adds:

> Good leadership is boundless. In addition to your academic triumphs, yours was also proved by complete victory over an all day massive electric power cut!

Although this would have been a major headache for many, to Sir Hermann this would not have posed a problem. He would have been able to carry on with or without the benefit of power. The fact that he clearly did this so successfully pays fitting tribute not only to the project itself, but also to his great skill as a communicator and a leader of men.

Sir Hermann, Connie the cat and a young friend relaxing in
the garden at Impington (1997)

THE THIRD RETIREMENT

Throughout all this period [1983-90] I have never neglected my scientific work. Being a theoretical person, all I need for it is pencil and paper. Even after I retire from the Mastership I expect this will keep me pretty busy. In my most energetic managerial days, the only time I had for my own research was effectively on aeroplanes and on trains. But I did manage a tolerable output of papers in all those years. One of my anecdotes about this period is of being stuck in an European airport, through the incompetence of an airline, for most of a day. When I published the paper resulting from my work there, I wondered whether to dedicate it to the airline in question![126]

Statute IV (Retirement and Resignation of the Master) requires the Master to retire 'on 31 July next following his or her seventieth birthday'. Sir Hermann celebrated his seventieth birthday on 1 November 1989 so retirement from the Mastership the following summer was inevitable. This would in fact be his third retirement, the first being his retirement from the Civil Service in 1979 at the age of 60 and the second being his retirement as Chairman of the Natural Environment Research Council just before his 65th birthday. Before his retirement, Vic Brown was the Conservationist in the Archives Centre. He recalls Sir Hermann speaking on the occasion of Vic's own retirement from the College, joking about the fact that this was only Vic's first retirement whereas he, Hermann, had done it three times and that he could thoroughly recommend it.

In the Master's Lodge, activities began to focus on the future; on a home for their retirement, on future academic and other pursuits to be enjoyed; on the transferral of papers to the Archives Centre; and on the third and final, retirement. Christine says that they made the decision to move from Surrey and settle in Cambridgeshire early in 1987. They sold the Reigate house in the autumn of 1987 and bought their house in Impington, just outside Cambridge, in April 1988. Thus in the months that followed items of furniture, books and other belongings could be moved to their new home at a relatively leisurely pace.

In January 1989 the attention of the College started to focus on the appointment of the next Master. A nominating committee was appointed, meetings held and six months later Alec Broers' appointment as the fourth Master of the College was confirmed. The appointment would take effect from 1 August 1990, so Statute IV would be satisfied and the Vice-Master's job done.

Churchill College is still a relatively young College, but fifty years is long enough to see the establishment of new traditions, one such tradition being the commissioning of a portrait of the retiring Master. Indeed, portraits of all the former Masters now hang proudly on the stairway leading up from the Senior Combination Room (SCR) to the Dining Hall and Fellows' Dining Room. The College is relatively open-minded about the artist and style of the portrait and the portraits are testament to this. The matter was discussed by the College Council on 8 November 1988:

Minute 11669 * Portrait of the Master (ref. Minute 11021)

In 1986 the Council had invited the Vice-Master, the Bursar and the Chairman of the Hanging Committee to consider the question of commissioning a portrait of the Master. Dr. Knott said that he had discussed the matter with the Bursar and Dr. Fraser and had obtained an estimate of the cost of a standard portrait in oils. Some members of the Council felt that a high-quality photograph would be preferable to a painting and it was agreed that Dr. Knott should consult the Master to see which he would prefer.

Following the meeting, the Vice-Master, John Knott, wrote to Sir Hermann:

I thought that it might be easier if I were to put down on paper one or two of the points raised at Council yesterday, so that you can mull them over at your leisure: if I were to try to put them verbally, it would constrain your thinking time.

The Council, of course, doesn't really like to make hard and fast decisions on anything, let alone a matter that involves artistic tastes and my impression of the "bottom line" is that it really boils down to what you would prefer.[127]

Although it is more than likely that the Council would have been perfectly capable of making a hard and fast decision on the matter on grounds of cost and suitability alone, it is quite certain that it did seek a steer from Sir Hermann. From John Knott's letter it becomes clear that there were general artistic concerns that portraits in oils could be rather dull and boring, but that in the main he was trying to convey that, within reason, the Council was prepared to do whatever Sir Hermann wished. Suggestions for alternative media included a high-quality photograph (as this might be a better way of bringing out features and character); or a drawing, or drawing and colour wash (for brightness of rendering and a modern look). However, in his letter, John Knott made one further suggestion:

Hologram: Full of exciting possibilities for a technological College and might be a 'first' for a Cambridge Master, but we have no idea either of cost or, more importantly, of the quality and permanence of the final product (would we need to have a spare photo in a drawer?) It might be worth thinking about however.[128]

It was indeed an exciting possibility. However, it did not fire Sir Hermann's enthusiasm and the idea was not pursued further. Instead the portrait artist and painter, June Mendoza, was commissioned to paint his portrait. She had been invited to do so at his suggestion as he had met her prior to coming to Churchill and she had told him then that she would like to do so one day. Thus it was that despite his heavy schedule, Sir Hermann made numerous trips to June Mendoza's studio in London during 1989 to sit for his portrait. Debbie Bondi recalls:

Hermann went for a 'sitting' on 2 October 1989 – slightly reluctantly I understand – because Christine was looking after an 18 month old Ben as I had gone into labour and was in the Rosie. When Claire was born about 11 a.m. that day, we rang Christine to tell her the good news. My recollection is that she telephoned Hermann at June Mendoza's to tell him about the arrival of the first granddaughter. However, she was slightly concerned that he would be a poor subject as he would be too excited to sit still and do what he was told!

It is clear that it was an experience greatly enjoyed by both Sir Hermann and the painter of his portrait. June Mendoza writes:

The portrait of Sir Hermann by the artist June Mendoza

I remember that Hermann was a delight to work with and that he was beautifully and professionally cooperative--- as someone who empathises instinctively with what it takes for someone else to execute and complete a particular work; in this case my portrait of him for the College.

On 26 November 1989 June Mendoza wrote to Sir Hermann and Lady Bondi following a visit to the College to display the finished portrait:

Thank you both once again for all your special care and for looking after me so beautifully.

Of course I am especially pleased that you, Christine, seemed to like our painting. That is always a breath holding moment when the wife confronts the work and on this occasion there followed a small procession of judges!

And to you, Hermann, my gratitude once again for your enormous cooperation and constantly pleasant humour throughout.[129]

Sir Hermann was moved to respond on 1 December 1989 as follows:

What a lovely letter you write! For me, it was a wonderful experience to see you working and the

Name detail on the door of Room 50A

final product is a triumph. We are all looking forward to seeing it here.

Perhaps there will then be another occasion to entice you up here![130]

Work on Sir Hermann's autobiography was also well under way by now. He had been approached by a publisher in 1986 to see if he would be interested in writing an academic autobiography. Although that particular publisher was ultimately unable to take the project further, by now Sir Hermann's appetite for such a project had been whetted and he set about finding a suitable publisher. In September 1988 he received confirmation that Pergamon Press was willing to publish and he immediately set to work. Small sections of the autobiography were typed by me and some others by Audrey Gawthrop, who was Anne Hammerton's secretary at the time, but the majority was dictated onto spools and sent to the publishers for transcription, being typed up by the typist with the aid of a Grundig 2002 Stenorette. Four Grundig 670 cassettes were purchased by Sir Hermann on 3 September 1988 at a cost of £3.39 each. A Pergamon Press contact report dated 17 November 1989 noted that:

> The final text will be ready by Christmas and the intention is to have the book published by 10 May 1990 to celebrate the 50th Anniversary of Winston Churchill's appointment as Prime Minister …

Sir Hermann put some thought into a possible title for his autobiography and he came up with five suggestions, some of which were based on things he had said. These were:

1. Better to be Lucky than Intelligent
2. Flying Roast Ducks

3. Oil and Space
4. Admiration and Internment
5. Science, Churchill and Me

Since these were his preferred titles, I think he might have been amused that I have chosen to use the second one as the title of this memoir. On completion of his autobiography, however, Sir Hermann received an advance on royalties of £500 less the costs of typing the manuscript, namely £303.45, so he earned himself the princely sum of £196.55 for his efforts. But that was never the point of the undertaking.

Working on his autobiography would inevitably give Sir Hermann cause to reflect on his past career, but it was also a signal that his third retirement was growing closer. The Master's Lodge was now beginning to focus on his departure, on moving his personal belongings to the new house in Impington, on transferring documents and papers to the Archives Centre and on transferring other working papers to 50A, the room that the College had allocated for his use following his retirement.

When I entered 50A some time in November 2005, I found a room full to the brim with papers, files, books and other items. These included files on the IBM Europe Science & Technology Prize, the Institute of Mathematics, the Social Morality Council, the World Economic Forum and the Royal Society. There was also a file entitled 'Master's MPhil Supervisions' and three entitled 'Conversations in Natural Philosophy: Bondi/Duncan'; as well as several for the Select Committee on Energy. Included among the masses of books in the room were many journals which were later to pose a major headache to the Librarian, Mary Kendall, in their re-housing. In addition to bound and unbound Proceedings, Year Books and Notes and Records of the Royal Society, there were unbound volumes of the Journal of Astrophysics and Astronomy, the IMA Journal of Mathematics and Monthly Notices of the Royal Astronomical Society, to name but a few. Others included:

Manchester Memoirs
Vol. 128 (1988/89) UNBOUND
Vols. 133-137 (1994/95 – 1998/99) UNBOUND
Vol. 140 (2001/02) UNBOUND
Index: 1781-1989 BOUND

as well as the Indian Academy of Sciences Yearbook (1997-2004). All these books and journals lined the walls and outer lobby of 50A. A typical teaching room in Churchill College, the room had an ill-matched assortment of furniture and seemed quite dated. But to me the crowning glory was the three drawer filing cabinet which was stuffed full of Agendas, minutes and papers for every one of the meetings Sir Hermann had attended since his retirement as Master. As these were unearthed, many bore the same tell-tale algebraic scribbles to which I had become so accustomed during his Mastership, signs that nothing had really changed.

In the summer of 1990 Sir Hermann's Mastership came to an end and he and Christine finally moved out of the Lodge. Alec Broers was admitted as fourth Master of Churchill College on 1 August 1990 and he and Mary moved into the Master's Lodge. Of the years that followed, Christine says:

After Hermann retired from Churchill he continued with many of his connections, such as the SRHE (Society for Research into Higher Education), Atlantic College, IMA (Institute of Mathematics and its Applications) and the RAS (Royal Astronomical Society). He also went regularly to conferences in Davos and to the Space Science Summer School in Alpbach. Without the help of a secretary he had to do some of the organising himself! So he stayed busy. We were back in India in late 1990 and again in 1995/6.

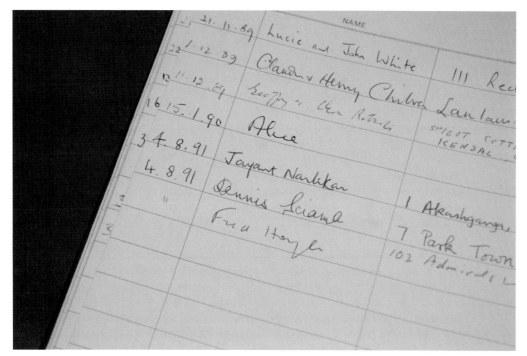

The Bondi's Visitors' Book

Nothing had changed.

Sir Hermann continued to be a presence in College throughout his third retirement, whether working in his College room, 50A, attending functions for both Fellows and staff, or giving keynote speeches at conferences and other events. Three such examples are illustrated below:

Beyond the Steady State Theory

At Sir Hermann's invitation, Wolfgang Rindler, a former By-Fellow, spent the academic year 1961/2 on sabbatical at King's College London. He writes:

> At that time the Steady State Theory of cosmology, which Bondi, Gold and Hoyle had invented in 1948, was in the last years of its huge popularity. But Hermann had already moved beyond it to the far deeper problems of gravitational radiation. I had the privilege of seeing Bondi, Gold and Hoyle together on two or three separate social occasions, the first time at the Master's Lodge of Churchill in about 1985. The chemistry of their friendship was very special, very touching and very mellow like an old wine. Each was a superb story teller, each had an exquisite sense of humour and a brilliantly quick intelligence. They could go on for hours with stories and jokes, feeding off each other. Hoyle's were usually the most risqué.

Christine's Visitors' Book records several visits by Tommy Gold and Fred Hoyle during Sir Hermann's time at Churchill College, the last recorded visit by Tommy Gold being to their Golden Wedding celebrations on 1

FLYING ROAST DUCKS

November 1997 and that by Fred Hoyle being on 4 August 1991. I can also remember a visit to the Lodge by Fred Hoyle. Dressed in shirt sleeves and with his broad Yorkshire accent, I remember the naturalness of the meeting and the complete lack of pretence. Why I should have expected otherwise I really don't know.

But any friendship can encounter difficulties and there are times when even friendship cannot over-ride academic propriety and the need to set the record straight. The Bondi-Gold-Hoyle friendship was no exception. Sorting out some papers at home in Impington recently, Christine found an exchange of correspondence between the three of them concerning claims that Fred Hoyle had made in his 1994 autobiography, 'Home is Where the Wind Blows'.[131] Tommy Gold had written to Sir Hermann on 14 May 1994:

> This is to tell you that Fred's book "Home is where the Wind Blows", has annoyed me greatly. Have you a copy of it? Around page 402 you will find that he claims that the beginning of the S-S Theory was his invention of the C-field... I know you described the story correctly, both at Cornell 14 years ago and also at the conference in Italy which I could not attend... I recall that he was already a little annoying in all the publicity that followed our papers, claiming then already more than his share. It was known as "Hoyle's Steady State Theory" after his popular lectures. [132]

Sir Hermann responded by fax on 23 May 1994:

> You are quite right: What Fred says in his autobiography is totally (and intentionally) misleading.... What should be done now? Your book will help (as my autobiography does). However, for the future historian, the differing personal stories of the chief actors are fun, but not a good way to establish the truth. ... In 1948 Fred's efforts to get his paper published before ours were undignified (to put it politely) and unsuccessful .... Fred's memory was always very selective, not to say selfserving. However, this totally wrong presentation is a bit much.

Together they discussed how best to deal with this difficulty and ever the diplomat, Sir Hermann was able to negotiate a compromise which was just about acceptable to all three. A joint statement setting the record straight was signed and submitted to *Nature* for publication on 5 January 1995.[133] The statement is beautifully scripted and the reader will be able to read both on the lines and between. But honour, dignity and academic integrity were preserved and the three, who were all to live into their eighties, maintained contact for the remainder of their lives.

By way of a postscript, a letter to Sir Hermann from Lady Barbara Hoyle dated 26 February 1990 beautifully describes their mutual raison d'être. She reported that 'Fred is well and very busy with his research work so all that is important is right with the world'.

Science, Churchill and Bondi

An electronic search on the College's administrative network in 2010 revealed a number of documents relating to Sir Hermann: from notice of a Post Prandial held in College after High Table on 5 March 1999 when Christine gave a talk on "Education and its fallacies: how one thing leads to another"; a formal letter of congratulations in February 2001 from the Vice-Master, Paul Richens, on the award of the Gold Medal of the Royal Astronomical Society; an exchange of correspondence in October 2001 with the current Bursar, Jennifer Brook, on Estates Committee matters; to the following document which was prepared on the occasion of Sir Hermann's 80th birthday on 1 November 1999. Since it reflects so well the affection and regard in which Sir Hermann was held, it is reproduced in full below:

On Thursday 18th November 1999 more than one hundred Fellows, their spouses and partners and close friends and family of Sir Hermann and Christine gathered in the Hall to celebrate the occasion. The whole arrangement was neatly choreographed by Archie Howie and Richard Keynes, Douglas Gough and Jack Miller spoke of Hermann's achievements and of his contribution to the College. Extracts from two of the speeches are printed below.

Archie spoke of Hermann's masterly way of dealing with the College Council and his astute and clever chairmanship of it. His ability to sum up is legendary, including at the end of a discussion not only the points which had been made, but several which had not been made all ready for the Minutes. He also paid tribute to Hermann by distributing a piece of apparatus called a rattleback which begins when spinning in one direction, halts, shuffles, and spins again in the opposite direction. An excellent model of a member of Council changing his mind.

Jack Miller recalled his own role in appointing Hermann as Master and how he had been in Swindon and asked a colleague of Hermann's for a view on him when Hermann was still a Chief Scientist. "He is quick, he is fast and a wonderful skier. He is always on the move so you have to be quick to catch him" was the verdict. Nepotism is always eschewed at Churchill, Jack noted, since during the period when the College was waiting for the new Master one of his children's application to enter Churchill was rejected!

The Master [Sir John Boyd] concluded the proceedings by pointing out that Hermann the astronomer had arranged a vast shower of meteorites to occur on his birthday, as indeed the Leo cluster appeared brightly on November 19th. With characteristic modesty Professor Bondi had arranged for cloud cover on that particular evening as well. John Boyd also referred to Hermann's immense intellectual contribution and the fun with which he entered into the deepest or lightest conversational subject.

A tribute also had to be made, and was made graciously, by the Master to Christine's wonderful support for Hermann and he mentioned her activities on a wider field within the community and in local teaching and through her support of charities. She was also the one person in the College remarkably well suited for dealing with the Visitor – The Duke of Edinburgh, whenever he comes.

In reply, Hermann spoke of his many friends and how when he had come he had been made to feel happier than almost any other place. He found the College friendly and with a very positive attitude. Following Archie's remarks he spoke of the rational way the Council made decisions and the loyalty with which individual members took those decisions. His view of the Governing Body was slightly more sanguine and he spoke with mock seriousness of the decisions of such a body being important but unpredictable. Lastly he told one of his famous stories which very few of us had heard before and which will not be printed here since it was a privileged communication to those who attended the occasion.

Richard Keynes: Since I must have known Hermann longer than anyone else here, may I give you some reminiscences of former days, for we first knew one another when we were in Trinity together in 1938. Then one day early in 1940 he suddenly disappeared from our midst, for in a short-lived panic that dangerous aliens of unproven loyalty were being harboured in the country, the government had hastily shipped a bunch of suspects off to Canada. But always efficient, Hermann managed to be accompanied on the exodus by his research supervisor, so that his mathematical researches were not interrupted, and he was elected to a junior research fellowship at Trinity in 1943.

Shortly afterwards, I too went down from Trinity and spent the next 2½ years working at the Anti-Submarine Experimental Establishment at Fairlie on the Clyde. In September 1942 I was involved in a successful plot to replace the Director of A/S.E.E., details of which are preserved in the Archives Centre, and to keep me out of further mischief I was transferred to the Admiralty Signals Establishment at Witley in Surrey to work on naval radar. And there to my surprise and pleasure one of the first people that I bumped into in the canteen was Hermann. Our paths there did not cross very closely, for I was designing the display systems for gunnery radars, while Hermann occupied an exalted place in the Theoretical Division, spending part of the time with a radar set near the top of Snowdon examining wave clutter in the neighbouring waters, and some of the rest theorizing with Fred Hoyle and Tommy Gold about continuous creation of the Universe. There was at one time talk of our all sharing a house in Surrey, but it never came to anything.

When the war was over, we moved back to Cambridge, me to belatedly take a degree in physiology, and Hermann to become notorious as the only person who could consistently make mathematics amusing. I suppose that his stories were well woven into his lectures.

In the 1970s, when Hermann was Chief Scientific Adviser to the Ministry of Defence, he invited me as an independent scientist with wartime experience both of radar and sonar, to serve on a high-powered Committee examining problems on Target Recognition. It was a most interesting experience for me to be allowed to see the remarkable advances that had been made since the relatively crude techniques that had been employed during the war. I was also impressed by the skill with which Hermann handled his Committee, and without giving offence politely punctured some of the scientific misconceptions of the Air Vice-Marshals, Admirals and Generals around him.

So I knew that he would have no difficulty in keeping our Governing Body under control, and was very happy when he was appointed to be our Master, and returned with Christine to Cambridge.

Douglas Gough: Hermann is most well known among young astrophysicists for the process of accretion of material onto massive astronomical objects, such as stars; the process bears his name: 'Bondi accretion'. Stars form from large gas clouds by spontaneous collapse; due to a localised compression. This result was first obtained by Sir James Jeans, but his analysis is valid only for the very initial stages. Hermann addressed what happens much later. It was possible to show that there is a maximum rate of materials accretion by the star which can never be exceeded; stars take time to grow. Hermann showed that most of the matter accretes behind the star in a long wake.

The process is controlled by the very complicated dynamics of the wake. Hermann made some amazing simplifying yet plausible assumptions, which enabled him to calculate how quickly the star is slowed down by the accreted material. This happens for the same reason as steam trains slow down when they take up water on the move. The outcome was a very simple formula which is still used today. This was one of Hermann's first major contributions to science, and it illustrates the manner in which Hermann has attacked complicated problems ever since, (both in science and administration): simplify the problem by removing obfuscation. If one can see clearly through the obfuscation, as Hermann can it is a remarkably successful technique to steer oneself safely through life.

Hermann's most significant astrophysical work was in the field of Relativity and Cosmology. He both made original discoveries and simplified the mathematical derivations of results already known. Some of his arguments are to be found in his truly beautiful book called simply 'Cosmology', published some 40 years ago. One such argument is 'the K calculus', based on a quantity which he called K, a measure of the ratio of what people travelling at different speeds would perceive to be the rate at which a clock ticks. It explains most elegantly and simply what has been called the twin paradox; that if one twin were to go off on a journey into space, on her or his return she or he would find that she or he is younger than her or his sibling. Despite the known successes of Einstein's theory of relativity, this was doubted by a substantial minority of scientists. The reason it was found difficult to comprehend is that the theory of relativity is based on the hypothesis that all 'observers' of nature must discover the same laws of physics. There is no such thing as an absolute observer; everything is relative. Indeed, Einstein was once heard to ask of a ticket collector: 'Does Heidelberg stop at this train?'. If an experiment were actually to be performed, the twins could compare their watches when they met up, and the relative ages of the two would then be certain. Hermann showed that the experiment is not necessary. What he did with his K calculus was to resolve the issue by pure intellect alone; he showed convincingly that it is indeed the space traveller who ends up being the younger, by an argument that is so clear and simple that I have had no difficulty whatever using it to explain the phenomenon to children in lectures at local secondary schools.

Cosmology is an intrinsically relativistic subject. Yet, as Hermann showed, the basic results can be obtained using Newtonian, rather than Einsteinian, relativity. This is the outcome of an ingenious application of the most basic of Einstein's relativistic ideas to the technically more straightforward three-century-old theory of Newton. I don't know whether it was Hermann who invented Newtonian cosmology in this form, but it was certainly he who made it accessible to the common student. Remarkably, the equations describing the evolution of the Universe are essentially the same as those derived from Einstein's equations, yet for the ordinary physicist-in-the-street they are much easier to derive, and superficially much easier to interpret in terms of everyday concepts.

Another subject on which Hermann has worked is gravitational radiation; it is the propagation of information by gravitational forces; gravity is not instantaneous, but propagates with the speed of light. It was in this context that I first encountered Hermann. He started by asking the question; 'If the Sun were suddenly to go out of existence, how long would it be before we on Earth would know it?' He then pointed out that this is not a question a scientist can even ask, because matter cannot simply disappear. But then he said, standing on the platform with his curly hair sticking out wildly on both sides of his head (somewhat more copiously than it does now) and grinning from ear to ear (as he still does); 'What if the Sun were to sprout horns?'. By then he had the complete attention of his entire audience, whom he had skilfully prepared to be receptive to the rest of his lecture, for this question is permitted by the laws of physics. Such a performance, I subsequently learned by experience, is typical of Hermann's lectures, which are memorable for years after.

One of the most beautiful, fascinating and controversial theories for which Hermann is (partially) responsible is the 'steady-state' theory of the Universe. According to Hubble's law, the Universe is expanding. The traditional, and currently accepted view is that the Universe exploded, essentially from a point, and has been expanding and thereby becoming more rarefied ever since. One of the obvious difficulties is the idea of a point in time before which there was no time.

Let me tell you how this was approached by Hermann and his colleagues, Tommy Gold and Fred Hoyle, with whom he was billeted during the war working on radar. The work was very intense, which meant that the trio had to spend at least 20% of their time on it. The rest of the time they devoted to a much more interesting activity: cosmology. Fred gave a lecture in Birmingham on the synthesis of the chemical elements, which explained how the elements have been created from hydrogen within

Sir Hermann and The Lady Soames, DBE, on the occasion of the celebration of thirty years of the French Government Fellowship Scheme (June 2004)

stars. The eminent physicist Rudolf Pierls asked: 'Where does the hydrogen come from?'. The next day the trio went to the cinema to see a new film called 'The dead of the night', a film which ended just as it had started, and on their way home Hermann said: 'I wonder if the Universe is like that?'.

This led to the ideas that the Universe is always like it was. Of course, there was the issue of Hubble's law to explain. That required that matter, in the form of hydrogen, had to be created out of nothing to compensate for the rarefaction produced by the expansion of the Universe. There was the answer to Pierl's question. However, many people found the idea to be unpalatable. The steady-state theory enjoyed a lease of life which lasted until the discovery of what has been interpreted as the remnant of the flash from the initial explosion – the so-called Big Bang. Most recently, evidence has been found that the Universe is evolving, at least locally, and is therefore not in a truly steady state. An evolving theory based on thinking similar to that associated with the original steady-state theory is being entertained by a few scientists; so a future reincarnation of the steady-state idea, no doubt in a rather different form, is not wholly out of the question.

There was a time during this productive wartime interlude when Hermann was visited by another Cambridge scientist, Raymond Lyttleton, who did not know the precise address. He stopped in the village and gave a description of the three scientists, and Hermann in particular, whereupon a villager's eyes lit up; he smiled and said: 'Ah! You mean the clever one!' As we all know, Hermann is clever not solely at mathematics, but I shall leave it to other Fellows to discuss his additional talents.

During this third retirement Sir Hermann continued to visit College regularly and to contribute where he could. On 21 June 2004 the College and the Scientific Service of the French Embassy in London celebrated thirty years of the French Fellowship at Churchill College as part of the Entente Cordiale centenary celebrations between this country and France. Of Sir Hermann's participation in this event, Professor Kelly wrote:

Sir Hermann Bondi spoke on "Science and Society" and [with a] particular emphasis on his time at the European Space Research Organisation, when facetiously he said that in order to get the public interested in the activities of the Organisation he had almost thought of planning a disaster in order to get attention. The Challenger disaster, the satellite which failed on launching, had given publicity in the USA. He spoke movingly of the accurate annunciation [*sic*] both of problems and of difficulties and solutions by French engineers, emphasising that what was important in a multi-national organisation like his, was engineering competence rather than nationality. He went on to say that in his opinion people regarded science quite wrongly as making ex cathedra statements whereas really science was more accustomed to disagreement than to complete agreement on any current problems.

He thought that it was the Newtonian precision in explaining the solar system which had given people this false impression and with characteristic originality pointed out that a much better example of the position of science is to think of weather forecasting. This is much more typical of the scientific method. Finally he gave a personal story of a development of measurement of the astronomical unit.[134]

LEFT: Sir Hermann and Lady Bondi at a staff party (2001)

REFLECTIONS

I have built so much of my personal life and of my career on being fit and untiring that the idea of being helpless and weak is most disagreeable to me.[135]

This memoir has observed Sir Hermann at work and at play and has also absorbed the views and comments of some of his colleagues and friends. But what was he like as a person? Those who knew him will remember his great capacity for telling jokes. He always had a funny story for every occasion, even when giving speeches or lecturing and I am sure many will remember the sound of his loud deep laugh echoing around a room. Roy MacLeod, a former Fellow, once travelled with him from Canberra to Sydney and says that he was a great person to sit next to on a flight. This is not hard to believe: never one to be quiet for long, Sir Hermann would have been capable of entertaining the entire flight.

Wolfgang Rindler, his former research student and colleague, has provided a summary with which many will concur:

On the human side, there are so many fine characteristics that stand out – his kindness, his brilliance, his wit, his humor, his phenomenal memory, his vast knowledge of geography, history and politics. I think the memory part is a big component of many outstanding scientists. It helps them bridge across known facts, to combine and build on past results of their own and of others. Even the sense of humor helps in science, not to take oneself or others unduly seriously. It leads to innovation, it prevents too much reverence for accepted results or authority. Einstein also had a big portion of it.

Outstanding about Hermann also was his enthusiasm and curiosity for new things. It inspired his zest for travel, often strenuous travel, all over the world. Linda and I had the good fortune to have Hermann and Christine stay with us for about a week here in Dallas some years ago – and their enthusiasm was electric! With them we saw so many sights and museums and gardens as we never would on our own!

And Hermann was a total gentleman – which is not always a given among great men. Only a few weeks ago I myself faced a serious ethical dilemma at work. As a real help (and not just because we have recently been thinking about him, she'd have said the same a year ago), Linda said to me: put yourself in Hermann's shoes, think what HE would have done in a case like this. And the thought really helped!

One of the College's first women Fellows was Pat Wright. She recounts the following:

Sir Hermann Bondi was exceedingly kind to me when a job move took me from Cambridge to Cardiff in 1998. Family commitments meant I needed to return to Cambridge on Friday evenings, spend the weekend and Monday here before returning to Cardiff on Monday evenings. I have a laptop and travel is easy; but I needed somewhere to work during Mondays that was free from domestic distractions. For a while I temporarily invaded any empty space in Churchill but when Hermann learned of the problem he was quick to offer his room in college as a regular pied-à-terre. Whenever I urged him to let me know if a date was inconvenient or if he would like to change the arrangement (i.e terminate it) he always replied that he was happy for the room to be used more and insisted that it made him feel less guilty about having a room in College. And so the arrangement continued until 2002, with Hermann dropping in from time to time to pick up mail and have a chat which was always accompanied by humorous reminiscences. He was a generous and delightful friend.

Another Fellow, Anny King, recalls:

I met Sir Herman only twice at Churchill. The first time was at the Churchill Debating Society where he came and participated fully in the students' debate. The second time was in the SCR prior to a special dinner. Both times, he struck me as being an exceptional person – not just because of his professional achievements which I knew were numerous, not just because he had been Master at Churchill for some years (I wasn't a Fellow then) and helped put Churchill (a new college at the time) on the Cambridge map and beyond, but because of his extraordinary humanity, simplicity and humility. He had an aura around him that made him stand out wherever he was. He would pay as much attention to the words of an eminent scientist as to those of a Fresher. He would treat you as an equal whoever you were. He would make you feel important (to him). He had the knack of making you feel happy once you'd exchange a few words or even a smile with him. There are encounters that stay with you forever and my two encounters with Sir Herman live on in my memory and whenever I recall these brief encounters I feel warm and happy inside. What a man! What a gentle man!

Sir Alcon Copisarow, a former Archives By-Fellow, recalls the following interaction with Sir Hermann which was so typical of his humour:

I was due to speak at an event in College. It was in May 2005, I think. Hermann was about to introduce me when he asked what he could say. I said "I don't know, what would you regard as significant?" He said to me, "I see you have already got eight fellowships to your name, how about 'for he's a jolly good fellow'?" I was rather doubtful about this and told him that I had stopped subscribing to most of those bodies. He said "Well you can't say you are an unqualified success then". We then went into the lecture theatre and he stood up and announced "we would have been content to have had a less distinguished speaker than Copisarow, but we couldn't find one.'

Of the student perspective, Paul Dilger, a former undergraduate student, wrote:

As Churchill's JCR President from '84/'85, along with my MCR opposite number and good friend Barry Hague, I had the privilege of working with Sir Hermann and benefiting from the occasional party at the Master's lodge.

He was quite simply the cleverest person I have ever met, with a prodigious intellect and yet possessing the rare ability to distil the most complex argument or concept into something blindingly clear and illustrative.

A gracious, skilled chairman of the College meetings and an elegant host, I never found him without an anecdote to illuminate any conversation or enhance a discussion.

His passing reminded me of how lucky we are as College students to brush shoulders with greatness before we make our own way in the world.[136]

Of those staff I have spoken to, many speak of Sir Hermann with great affection, of their appreciation of the fact that he would treat them as equals and of course, all recall his funny stories.

As Sir Hermann's autobiography neared completion, Mark Goldie, Fellow in History, was asked to read the text. On 29 November 1989 Mark responded with nearly three pages of constructive criticism. In it he queried:

> If I may say so, I find your story generous to a fault. That is to say, it has a virtue and a vice. You have never a bad word for anybody. There are of course libel laws! – yet it is plain that a spirit of equanimity and tolerance breathes through and it is very fine to see a memoir that does not seek to settle scores. The supreme generosity is your tolerance to a nation and a government that interned you. But at the same time, it seems strange that there seem never to be people whom you found anything but agreeable. There are hints of arguments at ESRO. But scarcely a sense of contention or conflict, or personality clash ever darkens the page. Can it have been so?[137]

Even the annoyance at Fred Hoyle's 'selective' account of the events leading up to the publication of the steady state theory, was just that: irritation at the misrepresentation of the facts at he knew them to be. He sought and achieved a diplomatic resolution in the interests of academic propriety, but it was never for personal gain or glory.

As a scientist, did Sir Hermann ever feel compromised by his administrative duties? In an interview in *Science Age* in 1986, he said:

> In my last 10 to 12 years as an academic scientist, I was involved in innumerable committees – making choices, decisions, etc. When I became an administrator, I always tried to do a little science and I think in these nearly 20 years I have managed to publish a paper every now and then. There must be about 20 papers I published in these 20 years. So, it hasn't gone down to nil. But, of course, one spends one's own time where one's interests lie, where one feels one can apply such abilities as one has to the best advantage. I have enjoyed all the things I have done. I often deeply regret there are only 24 hours in the day because there are so many things I'm interested in and would like to do. I haven't done them all. You've got to make your choice.[138]

But reflecting on his Mastership, what had he thought of this period of his life? Christine says that he thoroughly enjoyed his time as Master. Indeed, in his autobiography, he wrote:

> To have become Master of Churchill College was a particular pleasure for me, with my historical interests. To be the Head of the national memorial to such a great man who was so influential in my time and who [sic] I admired so greatly, was wonderfully to my taste. When I speak to our young people, at our Annual Founder's Day, I am only too conscious of my responsibility as being the last Master of Churchill College who experienced Winston Churchill's leadership as an adult. We have also greatly enjoyed our contact with Churchill's family, notably with his younger daughter, Mary Soames (whose much liked and admired husband Christopher unhappily died during our time at the College) and his grandson Winston Churchill MP.[139]

Sir Hermann was always proud of his robust health, taking the view that one only felt tired when one was bored. In his autobiography he recalls a routine medical examination.

> When one of the doctors finished explaining to me all the excellent features that he had discovered in my tests, he said "I am afraid the regulations don't allow me to issue certificates of immortality".[140]

Sir Hermann and Lady Bondi opening Golden Wedding presents at home in Impington (1997)

He never smoked, was apparently irritated at having to take anti-malarial pills when travelling overseas and whenever his back was playing up, found 'a combination of aspirin and red wine very helpful'.[141] In an article entitled 'Death' he wrote:

> That my own life will come to an end is something I have no quarrel with. To think of my death in the distant future evokes no particular feelings in me, to think of it in the relatively near future produces considerable feelings of serious irritation at all the jobs left unfinished, at all the experience I have gathered that will now be useless, at letting down all those to whom I am tied by love and friendship and all who relied on me for help and advice.[142]

When Sir Hermann discovered he was suffering from Parkinson's disease in 1996, in characteristic style he was determined that it would not stop him from doing the things he wanted to do. His daughter, Alice, says:

> He never complained, but tended to say things like "it's a bit of a bother". Of course, after the terrible fall in the May that left him unconscious for three days, things really weren't ever quite right again and his back caused him a lot of pain. I'd watch him trying to stand, clearly finding it agonising, but he would utter no more than "hmmmph".

Debbie Bondi agrees:

Hermann was determinedly cheerful despite the pain and the worry - he was always so pleased to see me or the children even when he was really uncomfortable.

Despite his failing state of health, Sir Hermann continued with his work. He had undertaken to write the biographical memoirs of Tommy Gold for the Royal Society but he now acknowledged that he was in need of help to complete this project. He asked Tim Cribb, the Tutor for Advanced Students, to find an applied mathematics/science PhD student, and a young Brazilian student called Yuri Sobral leapt at the chance. Yuri has lasting memories of the encounter:

Coming from a fluids dynamics background, and doing a PhD in fluid in DAMTP (Department of Applied Mathematics & Theoretical Physics), I sort of fitted the requirements, but I had never been in touch with cosmology and astronomy, and therefore was not familiar with Sir Hermann and his work. However, we agreed to meet at his house in Impington on 5 March 2005. I was a bit tense before the meeting: I had just arrived in the UK, my English was not perfect, I didn't really know what he wanted from me, and I was about to meet the former Master of a College. This was not helped by the fact that I arrived late because I was waiting for a taxi on the street and was not aware that in Cambridge you have to phone for them. But Lady Bondi answered the door with her characteristic gentle smile and I soon relaxed. We chatted in the living room for a while about my PhD work, my background, and his visit to Brazil. He then brought out a box full of notes and a list of articles he needed in order to complete Tommy Gold's biographical memoir. He was a bit concerned at how much time I would need to spend searching for the 36 or so articles he needed, but I told him it would be fine and that I was happy to help him. We agreed that I would return in a fortnight to check on progress. I remember that at the end of that first meeting Lady Bondi drove me back to the 'Pepperpots' (graduate houses) at Churchill, and on the way she told me stories of her time in Cambridge as a student, of how the University life worked for women in those days and of their time in the Master's Lodge at Churchill.

I returned to the house again on 20 March 2005 with copies of most of the papers Sir Hermann wanted. He was very focussed this time and after a brief chat about the articles he took me upstairs to his office. Due to his condition he had some trouble climbing the stairs. We started to separate the articles and order them according to subject and period. He stopped when he came to one particular article[143] and asked me what I thought of it. I was a bit embarrassed because I hadn't read all the articles. I had skimmed through many of them although I had tried to read this particular article as it had only three pages and no equations and I had expected that I would understand it. Unfortunately it had proved to be too far from my own sphere of understanding. So I just told him that it was "interesting". How immature was that! He stopped for a moment and looked at me and then we continued with our work. I really do think I disappointed him a bit with that comment. After we had finished, we went down to the kitchen and he told Lady Bondi "I have a lot of reading to do!" He was really motivated and had a lot of energy, despite his age and health, and I was impressed by his eagerness to complete the project.

I returned to his house a couple more times over the next two months that followed and we had some telephone contact whenever he needed extra help with a paper or whenever I found, or received from libraries or other universities, more articles, but I noticed that he was getting slower. His movements were more restricted and it was clear that it was taking him longer to climb the stairs to his office. The shaking was getting worse and once during a short period of quite strong shaking which was preventing him from reading a paper, he was clearly annoyed. Nevertheless I was still impressed that he was still so focussed on completing the memoir. A few months passed after that without news. One day I met Lady Bondi in the Porters' Lodge at Churchill and she told me that he was in hospital. She looked very tired and saddened but still had that very same warming 'grandma's' smile I remembered from the first time we had met. The next news I had of him was that he had sadly passed away. He was without doubt one of the greatest scientists and men I've ever met.

At Churchill College, Sir Hermann continued to attend College meetings right up until shortly before his death. His head would often start to shake, but he would carry on with what he was saying as if by doing so he could will it to stop. At other times his arm would start shaking badly, but I think that the shaking could be eased by the tablets. I also remember that he fell over in College once on his way to a Governing Body meeting and turned up in my office, his head bleeding. He sat quietly in my office for a few minutes and I made him a very weak cup of tea. But he was very soon back to his old self. Determinedly cheerful.

Christine says that Sir Hermann tended to be apologetic; worried that he was being a nuisance. He had been so proud of how fit and well he had been all of his life, but in the end he just gently deteriorated and didn't worry so much.

Sir Hermann died on 10 September 2005. A farewell ceremony was held on 19 September at 12 noon at the Cambridge Crematorium. The ceremony, a humanist, non-religious ceremony, was conducted by Helen Bush of the Independent Association of Humanist Celebrants. During the ceremony one of his three daughters, Liz Bondi, gave readings from the poetic work Man, written by his long-standing friend, the poet Ronnie Duncan, in 1970.[144]

Following Sir Hermann's death it was decided to hold a concert in his honour in Churchill College on 26 November 2005. The programme included Mozart's Clarinet Concerto in A major, K.622, Sibelius' Karelia Suite, op.11 and Corelli's Concerto gross op.6 no. 8. It also included some suggestions from Christine on what Sir Hermann would have talked about if he had been invited to appear on 'Desert Island Discs':

"Here are one or two thoughts about what Hermann would have wanted to mention on 'Desert Island Discs'. He was always a keen climber, especially in the Alps and in late life would be very proud of more mundane climbing when he did it better or faster than anyone else. He was particularly proud in April 2002 when he got to the top of a pagoda which nobody else tried and sauntered back to the rest of us very casually.

He was always fascinated by maps and could frequently be found tracing not only his own but other people's expeditions on maps spread all over the floor. And of course guide books, particularly Baedekers.

Alice [one of Hermann's daughters] recalls innumerable visits to art galleries when Hermann went straight to portraits, especially those of old men of real character, so Rembrandt was very popular – also Holbein. And he loved to look at those in the Hall at Trinity".

For his book, Christine suggests that Hermann might have chosen Gibbon's 'Decline and Fall'.

As for music, his classic choices were 'Schubert's Unfinished' and Mozart (almost anything but Alice mentions 'Eine Kleine Nachtmusik').

And innumerable letters referred to his enjoyment of playing with children and their toys![145]

But for Churchill College's third Master, I am sure that both Churchill College and its Founder would concur with Lord Carrington's summation in his foreword to Sir Hermann's autobiography:

Churchill College has been fortunate to have him as its Master. I am sure that the Great man after whom it was named would have looked with approval on his tenure of office and been gratified that so distinguished a man should be its Master.[146]

A quiet moment (1980)

APPENDIX 1
CONDENSED BIOGRAPHY

Hermann Bondi, FRS (1959), KCB (1973); born 1 November 1919, Vienna; son of Samuel and Helene Bondi; educated at the Realgymnasium, Vienna; undergraduate, Trinity College, Cambridge, 1937-40; interned, 1940-1; Admiralty radar research, 1941-4; Research Fellow, Trinity, 1943-9, Fellow, 1952-4; naturalised Briton, 1946; University Lecturer, Cambridge, 1945-54; Professor of Mathematics, King's College, London, 1954-71; Director General of the European Space Research Organisation, 1967-71; chief scientific adviser, Ministry of Defence, 1971-7; chief scientific adviser, Department of Energy, 1977-80; chairman, National Environmental Research Council, 1980-4; Master, Churchill College, Cambridge, 1983-90; secretary, Royal Astronomical Society, 1956-64; president, British Humanist Association, 1982-99; honorary degrees from Bath, Birmingham, Plymouth, Portsmouth, St Andrews, Salford, Southampton, Surrey, Sussex, Vienna, York; Einstein gold medal, 1983; Royal Astronomical Society gold medal, 2001; married Christine Stockman, 1947; three daughters, two sons; died 10 September 2005, Cambridge; author of *Cosmology* (1952), *The Universe at Large* (1961), *Relativity and Common Sense* (1964), *Assumption and Myth in Physical Theory* (with Kathleen Ollerenshaw, 1967); scientific papers in *Monthly Notices of the Royal Astronomical Society*, *Proceedings of the Royal Society*, *Proceedings of the Cambridge Philosophical Society*; autobiography: Science, *Churchill and Me* (1990).[147]

APPENDIX 2
CONTRIBUTORS

The following, who are all cited in the text, responded to a request for reminiscences and anecdotes about Sir Hermann Bondi. I am extremely grateful to them for their contributions. (See Index of Names for further information).

Allen, Mr M J A
Barnett, Mr C
Beveridge, Miss M M
Bondi, Ms A (Alice)
Bondi, Lady C M (Christine)
Bondi, Ms D (Debbie)
Brook (formerly Rigby), Mrs J M
Brown, Mr V
Copisarow, Sir Alcon
Cribb, Mr T J L
Dilger, P (for *Churchill Review*)
Finch, Professor A M
George, Mr H
Goldie, Dr M A
Hammerton, Miss A E
Livesley, Dr R K
MacLeod, Professor R
Mendoza, Ms J
Miller, Dr J A *
Miller, Mrs M
Parker, Dr J
Rindler, Professor W
Smick, Mr C W
Sobral, Dr Y D
Stephenson, Professor J
Tandan, Dr B
Tester, Dr M A
Wright, Professor P A

* by kind permission of Mrs Marcia Miller

APPENDIX 3
INDEX OF NAMES

BF   By-Fellow
F   Fellow
M   Master
G   Postgraduate
U   Undergraduate
S   Staff

Adkins, Dr T (Tess)	Fellow of King's College
Adrian, Lord (Richard)	F 1960-95
Allen, M J A (Michael)	F 1985-; Bursar 1990-1998
Andrews, Mr N H (Neil)	F 1983-1995
Atiyah, J, Mr J (John)	S 1991-95; Computer Systems Manager
Barnett, Mr C (Correlli)	F 1976-; Keeper of the Archives 1977-95
Bates, Mr J (Jonathan)	Sir Hermann's Private Secretary at NERC
Beveridge, Miss M M (Mary)	S 1972-91; College Secretary, then Registrar
Bondi, Ms A (Alice)	Eldest daughter
Bondi, Lady C (Christine)	Widow of Sir Hermann Bondi
Bondi, Ms D (Debbie)	Youngest daughter
Bondi, Sir Hermann	M 1983-90; F 1990-95
Broers, Lord (Alec)	M 1990-96; F 1996-
Brook, Mrs J M (Jennifer) formerly Rigby	F 1999 -; Bursar 1999-
Brown, Mr S H (Sid)	S 1962-89; Chief Maintenance Engineer
Brown, Mr V (Vic)	S 1970-98: Conservationist
Bullock, Mr P (Peter)	S 1979-99: Head Porter
Campbell, Dr R C (Colin)	F 1962-; Senior Tutor 1975-85
Carberry, Mr J (Joseph)	S 1984-07; Dining Hall Manager
Clements, M H (Martin)	U 1980-83
Copisarow, Sir Alcon	Archives BF 2005
Cribb, Mr T J L (Tim)	F 1969-; Tutor for Advanced Students 2001-06
Dilger, Mr P J (Paul)	U 1983-86

Dixon, Dr W G (Graham)	F 1964-; Tutor for Rooms/Tutorial Bursar 1978-03
ESRO	European Space Research Organisation
Finch, Professor A M (Alison)	F 1972-93; F 2003-; Vice-Master 2005-06 and 2008-2012
Gawthrop, Mrs A G (Audrey)	S 1982-04; Bursar's Secretary, then Secretary to Conference Manager
George, Mr H (Hywel)	F 1972-; Bursar 1971-90
Goldie, Dr M A (Mark)	F 1979-; Vice-Master 1993- 99
Gough, Professor D O (Douglas)	F 1972-; Tutor for Rooms & Finance 1974-77
Hague, Mr B (Barry)	U 1980-83; PG 1983-88
Halson, Mrs P (Paula)	S 1988-; BF 2006-; Master's Secretary 1988-91; Registrar 1991-
Hammerton, Miss A E (Anne)	S 1974-91; Domestic & Conference Manager
Hawthorne, Sir William	M 1968-83; F 1984-2011
Hayden, Mr M (Martin)	S 1975-; Head Chef
Hoskin, Dr M A (Michael)	F 1969-; Fellows' Steward 1987-88/1992-93; President of the SCR 1981-91
Howie, Professor A (Archie)	F 1960-; Dean 1963-5; President of the SCR 1999-10
Kapitza, Professor P (Pyotr)	Honorary Fellow 1976-84
Kendall, Miss M I (Mary)	S 1984-00; F 2001-; Librarian
Kelly, Professor A (Anthony)	F 1960-67; 1985-;
Keynes, Professor R D (Richard)	F 1960-2010
King, Mrs A N (Anny)	F 1994-
King, Dr F H (Frank)	F 1977-
Knott, Professor J F (John)	F 1967-06; Vice-Master 1988-90
Livesley, Dr R K (Ken)	F 1960-
MacLeod, Professor R (Roy)	F 1966-1970
Mee, Mr R M (Richard)	S 1976-; Deputy Head Chef
Mendoza, Ms J (June)	Portrait Artist and Painter
Miller, Dr J A (Jack)	F 1961-07; Vice-Master 1978-88
Miller, Mrs M (Marcia)	Widow of Jack Miller; Christina Kelly Associate
NERC	Natural Environment Research Council
Northeast, Miss C H (Christine)	F 1991-2004
Packwood, Mr A G (Allen)	S 1995-02; F 2002-; Senior Archivist, then Director of the Archives Centre
Parker, Dr J (Jonathan)	G 1987-92; MCR President 1988-89
Pledger, Mrs B (Brenda)	S 1984-1989; Bedmaker/Master's Lodge Housekeeper
Pledger, Mr G P (Graham)	S 1964-; Groundsman, then Deputy Head of Grounds & Gardens
Pole, Professor J R (Jack)	F 1963-1979; Vice-Master 1975-78
Richens, Professor P (Paul)	F 1994-2005; Vice-Master 2000-05
Rindler, Professor W (Wolfgang)	BF 1989
Roxburgh, Professor I W (Ian)	G 1960-63
Smick, Mr C J (Chris)	BF 1993
Soames, Lady (Mary)	Honorary Fellow 1983-

Sobral, Dr Y D (Yuri)   G 2004-08

Stephenson, Professor J (John)   National Director, RSA HEC Project 1988-98

Stephenson, Mr M I (Malcolm)   S 1982-07; Butler/Buttery & Cellar Manager

Tandan, Dr B (Banmali)   G 1975-78; BF 2003-05

Tester, Professor M (Mark)   F 1988-90; 1994-03

Tizard, Mr R H (Dick)   F 1960-05; Senior Tutor 1966-75

Todd, Rt Hon Lord (Alexander)   Honorary Fellow 1971-97

Tristram, Dr A G (Andrew)   F 1967-; Senior Tutor 1985-92

Walker, Mrs E (Evelyn)   S 1975-06; Dining Hall Manager

Westwood, Dr B A (Brian)   F 1966-; Wine Steward

Williams, Mr K R (Keith)   F 1983-07

Wright, Professor P A   F 1972-97

APPENDIX 4
BIBLIOGRAPHY

Bondi, H (1981) 'Jobs I have Enjoyed', *Mathematics and its Applications*, Vol 17, August/September

Bondi, H (1982) 'Why Science must go under the Microsope', *Times Educational Supplement*, 10 September 1982

Bondi, H (1986) 'Hermann Bondi: Science, Religion and Star Wars …', *Science Age*, Vol 4, No. 8, Bombay: Nehru Centre

Bondi, C M and Bondi H (1989) 'Traffic Jams', *SCOPE*, Issue 18, Winter

Bondi, H (1990) 'Reflections on Higher Education for Capability', RSA: *Higher Education for Capability Update:* Issue 2, February

Bondi, H (1980) *Relativity & Common Sense: a new Approach to Einstein*; unabridged, corrected reproduction of original (1964) edition; Dover

Bondi, H (1990) *Science, Churchill & Me: the Autobiography of Hermann Bondi*, Exeter: Pergamon

Briggs, A, ed-in-chief (1989), *Longman Encyclopaedia*, Longman

Churchill Archives Centre, The Papers of Sir Herman Bondi: online catalogue at: http://janus.lib.cam.ac.uk/db/node.xsp?id=EAD%2FGBR%2F0014%2FBOND

Churchill Review (2004), Vol. 41

Churchill Review (2006), Vol. 43

Duncan, R (1970) *Man: The Complete Cantos*, London: Rebel Press

Goldie, M (2009) *Churchill College, Cambridge: The Guide*, Cambridge: Churchill College

Goldie, M (2005) Expanded version of the talk given by Mark Goldie, Fellow and former Vice-Master, at Sir Hermann's Memorial Concert on 26 November 2005

Higher Education for Capability Archives, http://www.heacademy.ac.uk/heca (accessed 26 October 2009)

Howse, D (1993), *Radar at Sea*, Macmillan

Hoyle, F (1994), *Home is where the Wind Blows: Chapters from a Cosmologist's Life*, University Science Books

Neuberger, J and White J A (1991), *A Necessary End: Attitudes to Death*, London: Macmillan

Ollerenshaw, Dame Kathleen: Biography: http://www-history.mcs.st-and.ac.uk/Biographies/Ollerenshaw.html, accessed 24 November 2009

Ollerenshaw, K and Bondi, B (1982), *Magic Squares of Order Four*, Philosophical Transactions of the Royal Society of London

Ollerenshaw, K (1986), *On 'most perfect' or 'complete' 8 x 8 pandiagonal magic square'*, Proc. R. Soc. London A 407, 259-281

Roxburgh, I W (2007), *Biographical Memoirs of Fellows of the Royal Society*, Vol. 53, London: RSA

Statutes, Ordinances and Regulations of Churchill College, Cambridge

Churchill Archives Centre: Bondi Papers:

BOND Acc 1470 Box 3: Cambridge Executive Education Programme

BOND 1/10: Mastership

BOND 1/15: Master's Appointment and First Year

BOND 1/16: Miscellaneous College Correspondence – 1983-87

BOND 1/17: Miscellaneous College Correspondence – 1988-89

BOND 3/13: Miscellaneous Correspondence – 1958-86

BOND 3/17: Miscellaneous Correspondence – 1986-87

BOND 3/20: Miscellaneous Correspondence – Sept 1989 – Jan 1990

BOND 18/1: Naval Radar

BOND 18/2: Admin of Naval Radar Trust

Churchill Archives Centre: College Archive:

Buildings Committee Minutes: 1985

CCAR 101/5/15: Sir William Hawthorne/Swimming Pool 1979-81

CCAR 402/1: Master's Lodge 1963-81, 1990-97

CCRF/116/33: Obituaries – Hermann Bondi

College Council Minutes: 1978, 1984, 1988, 1989

College Council Paper: CC/98/83

Private Papers:

Items in the list below without reference numbers were recently given to the Churchill Archives Centre following completion of this memoir and are currently being catalogued. Please contact Churchill Archives Centre for further information.

Autobiography (original typescripts and correspondence)

Bondi/Gold: Hoyle Autobiography (BOND 3/25)

Cambridge Executive Education Programme (BOND 1/42)

Churchill College 1983-90 (BOND 1/34)

Dame Kathleen Ollerenshaw and Sir Hermann Bondi, FRS, Magic Squares of Order Four. Typescript (original text) December 1981

Enumeration for all Most-perfect Squares of Order n=p2 r (first draft), Dame Kathleen Ollerenshaw

Correspondence between Sir Hermann and Jonathan Parker 1988-1999

General Correspondence 1989/95 (BOND 3/40)

Letters to Newspapers (BOND 2/9A)

List of Journals

Mastership 1982 -1992 (BOND 1/32)

Naval Radar

RSA and Higher Education for Capability

Typescript of paper entitled 'The Origins of the Steady State Theory', Sir Hermann Bondi, FRS, June 1990 (BOND 2/18)

Photo References:

Documents and Images at pp. 16 (CCPH/3/5/2), 23 (CCPH/4/17), 26 (BOND 1/10), 27 (BOND 2/7) and 76 (BOND/18/2A) are held by the Churchill Archives Centre.

APPENDIX 5
END NOTES

1 Roxburgh, I W, Scientific Memoirs of Fellows of the Royal Society, Vol. 53, 2007

2 Expanded version of the talk given by Mark Goldie, Fellow and former Vice-Master, at Sir Hermann's Memorial Concert on 26 November 2005 (CCRF/116/33)

3 Bondi, H, Science, Churchill & Me, p. 3

4 Bondi, H, ibid, p. 5

5 Bondi, H, ibid, p. 21

6 Bondi, H, ibid, p. 31

7 Bondi, H, ibid, p. 41

8 Bondi, H, ibid, p. 50

9 Goldie, M, 'Third Master of Churchill College: expanded version of a talk given at a memorial concert on 26 November 2005', p. 11 (CCRF/116/33)

10 Private Papers: Typescript of paper entitled 'The Origins of the Steady State Theory', Sir Hermann Bondi, FRS, June 1990, pp. 4-5

11 Ibid, pp. 6-75

12 Bondi, H, 'Jobs I have Enjoyed', Mathematics and its Applications, pp. 164-165

13 Bondi, H, Science, Churchill & Me, pl 80

14 Goldie, M, 'Third Master of Churchill College: expanded version of a talk given at a memorial concert on 26 November 2005', pp. 3-4 (CCRF/116/33)

15 Bondi, H, Science, Churchill & Me, pp. 93-4

16 Ibid, p. 104

17 Ibid

18 Ibid, p. 109

19 In Roxburgh, I W (2007), Scientific Memoirs of Fellows of the Royal Society, Vol. 53, London: RSA

20 Bondi, H, 'Jobs I have Enjoyed', Mathematics and its Applications, p. 166

21 In Roxburgh, ibid.

22 Bondi, H, 'Jobs I have Enjoyed', Mathematics and its Applications, p. 166

23 Goldie, M, 'Third Master of Churchill College: expanded version of a talk given at a memorial concert on 26 November 2005', p. 4

24 Bondi, H, 'Jobs I have Enjoyed', Mathematics and its Applications, p. 166

25 I am referring to the archives of Churchill College which are also housed in the Churchill Archives Centre.

26 Churchill College Archives: confidential papers

27 Churchill College Archives: confidential papers

28 Bondi, H, Science, Churchill & Me, p. 122

29 Ibid, pp. 122-3

30 Private Papers: Mastership 1982-92

31 Private Papers: ibid

32 BOND 1/15

33 BOND 1/10

34 Private Papers: Mastership 1982-92

35 BOND 1/15

36 BOND 1/15

37 Statutes, Ordinances and Regulations of Churchill College, Statute III

38 Bondi, H, 'Jobs I have Enjoyed', Mathematics and its Applications, p. 166

39 Statutes, Ordinances and Regulations of Churchill College, Statute III.3

40 Bondi, H, 'Jobs I have Enjoyed', *Mathematics and its Applications*, p. 165

41 BOND 1/15

42 Private Papers: Mastership 1982-92

43 Churchill College Archives: College Council Minutes: 14 February 1984

44 Bondi, H, 'Jobs I have Enjoyed', *Mathematics and its Applications*, p. 165

45 Quoted by Paula Halson in Chapter 3

46 BOND 1/15

47 Churchill College Archives: College Council Paper: CC/98/83

48 Churchill College Archives: Master's Lodge

49 Goldie, M, Churchill College, Cambridge: The Guide, p. 36

50 Private Papers: List of Journals

51 Christine Bondi thinks this was a gift to Sir Hermann's father from a grateful patient although she does not know whether the patient was the artist or not.

52 Graham is the longest serving member of staff. He joined the College in March 1964 and is still in post.

53 Private Papers: Churchill College 1983-90

54 Bondi, H, 'Jobs I have Enjoyed', *Mathematics and its Applications*, p. 164

55 Private Papers: Churchill College 1983-90

56 BOND 1/16 Part II

57 Private Papers: Churchill College 1983-90

58 Churchill College Archives: Buildings Committee Minutes 1985

59 Private Papers: Churchill College 1983-90

60 Private Papers: Churchill College 1983-90

61 CCAR 101/5/15

62 Ibid

63 Ibid

64 Churchill College Archives: Buildings Committee Minutes:1985

65 BOND 1/16

66 Marcia recalls that this was the annual IBM conference dinner which was held in the Long Vacation

67 BOND 3/13

68 Private Papers: Churchill College 1983-90

69 Private Papers: Exchange of correspondence between Sir Hermann and Jonathan Parker: January – March 1989

70 BOND 1/16

71 BOND 1/16

72 Bondi, H, 'Hermann Bondi: Science, Religion and Star Wars …', *Science Age*, p. 43

73 Private Papers: RSA and Higher Education for Capability

74 Bondi, H, *Science, Churchill & Me*, p. 133

75 Private Papers: Churchill College 1983-90

76 BOND 3/17

77 BOND 1/17B

78 Private Papers: General Correspondence 1989-95

79 Private Papers: General Correspondence 1989-95

80 Private Papers: General Correspondence 1989-95

81 BOND 3/17

82 BOND 1/17B

83 Private Papers: General Correspondence 1989/95

84 BOND 3/19

85 Private Papers: General Correspondence 1989-95

86 Bondi, H, 'Jobs I have Enjoyed', *Mathematics and its Applications*, p. 164

87 Bondi, H, *Science, Churchill & Me*, p. 133

88 Letter to the Guardian, 14 May 1986, in Private Papers: Letters to Newspapers

89 Bondi, H, 'Why Science must go under the Microsope', *Times Educational Supplement*, 10 September 1982

90 Ibid

91 BOND 1/16 Part II

92 Bondi C M and Bondi H, 'Traffic Jams', SCOPE, Winter 1989, pp 4-7

93 BOND 1/10

94 BOND 1/10

95 BOND Acc 1470 Box 3

96 Ibid

97 Private Papers: Cambridge Executive Education Programme

98 Ibid

99 Private Papers: RSA and Higher Education for Capability

100 Ibid

101 Ibid

102 Bondi, H, 'Reflections on Higher Education for Capability', RSA Higher Education for Capability Update, Issue 2, February 1990, p 1

103 Higher Education for Capability Archives, http://www.heacademy.ac.uk/heca (accessed 26 October 2009)

104 Bondi, H, 'Hermann Bondi: Science, Religion and Star Wars …', Science Age, Vol 4, 8 November 1986, p 41

105 Letter to the Times Educational Supplement, 21 November 1988, in Private Papers: Letters to Newspapers

106 Longman Encyclopaedia, published 1989

107 A Chinese proverb

108 Bondi, H, Science, Churchill & Me, Preface

109 Churchill College Archives: Council Minutes 1988

110 Bondi, H, Science, Churchill & Me, p. 129

111 See Memoir of Dame Kathleen Ollerenshaw: http://www-history.mcs.st-and.ac.uk/ Biographies/Ollerenshaw.html, accessed 24 November 2009

112 The folder is entitled Magic Squares of Order Four, Dame Kathleen Ollerenshaw and Sir Hermann Bondi, FRS

113 Ollerenshaw, K: On 'most perfect' or 'complete' 8 x 8 pandiagonal magic square', Proc. R. Soc. London A 407, 259-281 (1986)

114 Private Papers: Enumeration for all Most-Perfect Squares of Order n = p2r (First Draft)

115 Quote taken from Nicholas Ruyter, Journal of Travels in Cathay and the Eastern Indies, 1683, in Private Papers: Naval Radar

116 Howse, D, Radar at Sea, Macmillan, p. xv

117 Private Papers: Naval Radar

118 Ibid

119 Ibid

120 Ibid

121 Ibid

122 Ibid

123 BOND 18/1

124 Private Papers: Naval Radar

125 Ibid

126 Bondi, H, Science, Churchill & Me, p. 126

127 Private Papers: Mastership

128 Ibid

129 BOND 3/20

130 Ibid

131 Hoyle, F (1994) Home is Where the Wind Blows: Chapters from a Cosmologist's Life, University Science Books

132 Private Papers: Bondi/Gold: Hoyle Autobiography

133 In Nature, Vol 373, 5 January 1995: http://www.nature.com/nature/journal/v373/n6509/p df/373010b0.pdf, accessed 19 February 2010

134 Professor Anthony Kelly writing in the Churchill Review (2004), Vol. 41, p 46

135 Typescript of article entitled 'Death', written for book published by Rabbi Julia Neuberger and Revd Canon John White in 1991 (Macmillan), in Private Papers: General Correspondence 1989-95

136 Churchill Review (2006), Vol. 43, p. 23

137 Private Papers: Autobiography

138 Bondi, H, 'Hermann Bondi: Science, Religion and Star Wars …', Science Age, p. 47

139 Bondi, H, Science, Churchill & Me, pp. 123-124

140 Ibid, p. 129

141 Ibid, p. 129

142 Typescript of article entitled 'Death', written for book published by Rabbi Julia Neuberger and Revd Canon John White in 1991 (Macmillan), in Private Papers: General Correspondence 1989-95

143 The article to which Yuri Sobral was referring was 'Energy Transfer by gravitation in Newtonian Theory', Bondi, H, McCrea, W H, 1950

144 Duncan, R, The Complete Cantos, London, Rebel Press

145 From the Programme for the Concert to Celebrate the Life of Hermann Bondi, on 26 November 2005

146 Bondi, H, Science, Churchill & Me, Foreword

147 Expanded version of the talk given by Mark Goldie, Fellow and former Vice-Master, at Sir Hermann's Memorial Concert on 26 November 2005. (CCRF/116/33)